Uncertainty Analysis for Engineers

Vincent W. Uhl and Walter E. Lowthian, editors

Paul R. Dunlap
John G. Frith
Alan Gleit
C-S Kiang
Ralph H. Kummler
Steven T. Sullivan
Henry C. Thorne
Vincent W. Uhl
Charels H. White
R. Gordon Winklepleck

AIChE Symposium Series

Number 220 1982 Volume 78

Published by

American Institute of Chemical Engineers

345 East 47 Street New York, New York 10017

American Institute of Chemical Engineers
345 East 47 Street, New York, N.Y. 10017

AIChE shall not be responsible for statements or opinions advanced in papers or printed in its publications.

ISBN 0-8169-0244-5

Library of Congress Cataloging in Publication Data
Main entry under title:

Uncertainty analysis for engineers.

(AIChE symposium series; no. 220)
Includes bibliographical references.
1. Engineering—Management—Congresses. 2. Risk management—Congresses. 3. System analysis—Congresses. I. Uhl, Vincent W. II Lowthian, Walter E., 1943- . III. Series
TA190.U53 1982 658.4'03 82-24443
ISBN 0-8169-0244-5

Printed in the United States of America by
Twin Production & Design

FOREWORD

This series of papers demonstrates that method for uncertainty analysis have been established and are being used for a range of situations in industry. In the past for engineers the emphasis has been on applications related to plant economic feasibility. The papers in this volume serve to show that uncertainly and risk analysis are being applied to a range of appropriate situations in technology. The contributions not only deal with economics feasibility, but also are concerned with system design, operations analysis, and the evaluation of test data. Now that engineers and scientists are becoming conversant with the available techniques, their application can be expected to grow rapidly in the near future.

Vincent W. Uhl
Walter E. Lowthian

CONTENTS

UNCERTAINTY ANALYSIS AND RISK ASSESSMENT FOR MANAGEMENT

Evaluations of completed projects frequently differ significantly from the original estimate. To show better the likely uncertainty in any new investment proposal, some form of risk analysis should be prepared. Management attitudes towards risk and the pro's and con's, and typical applications of three common techniques—sensitivity analysis, decision trees, and Monte Carlo analysis—are discussed.

HENRY C. THORNE
Amoco Chemicals Corporation
Chicago, ILL.

We live in a complex and uncertain world, and risk is always with us. Probably the most certain thing is that the future will be different from what we expect. This difference is often significant, and can have a major positive or negative impact on the profitability of investments.

Different types of projects also differ substantially in terms of risk. High risk projects include such activities as entry into completely new markets, utilization of a radically new process, most acquisitions and joint ventures. Relatively low risk projects include expansion of existing businesses, debottlenecking existing plants, and many cost-saving projects.

Expected return on investment is not always proportional to risk. The most ideal combination is a low-risk project with a high rate of return. Cost-savings projects are good examples of this. Here, the key marketing variables are much less significant, and rates of return of 50 per cent or more are common. However, a high rate of return for a proposed change is not necessarily a plaudit for management--it may simply indicate the relative inefficiency of past operations.

WHAT UNCERTAINTY CAN BE EXPECTED?

Very little appears to have been published to show the measured statistical uncertainty in profitability estimates, based on performance results, although much has been written on the importance of and techniques for post-installation appraisals.[1] A report on three Canadian projects[2] demonstrated the importance of good marketing estimates, but was not extensive enough to provide good ground rules on the variability to be expected.

More recently, information on 40 large investments was reported.[3] Most of the performance estimates were based on two to five years of actual operations, as it was not practical to wait the usual 15 to 20 years to project termination. The projects were analyzed by the discounted cash flow rate of return method, noting: (1) inaccuracies in volume estimates, (2) margin and price uncertainties, and (3) a residual category called "all other" which includes investment, timing, and other special effects.

In general, the margin and price forecasts were found to be optimistic. The other variables showed significant scatter but averaged close to the original estimates. Price data were much more limited than margin information, but from the data available price appears to be the primary factor causing the changed margins.

0065-8812-82-6124-0220-$2.00

A statistical analysis provided a useful estimate of the quantiative variability in profitability caused by each of the three factors studied. The results are summarized in Table 1.

The smoothed distributions for the four categories are shown in Figures 1 through 4. Interestingly, only the margin distribution is drastically skewed from the normal bell-shaped curve. The skewness of the total effect, shown in Figure 4, is much reduced because of the influence of the volume and "all other" variables. Also, the means of the volume effect, and the "all other" effect are essentially zero, indicating that on the average the individual estimates were well made. Again, on the average, the downward bias in actual profitability versus original estimate appears to be mainly due to an overly optimistic margin estimate. This in turn is probably the result of greater competitive pressures than anticipated.

Despite good average attainment of all estimates but margin, significant variation in each category can occur for individual projects. Much of this will probably always remain, as the greatest variability is in the marketing area.

For a normal distribution, the standard deviations shown in Table 1 include slightly over 68 per cent of all results. Although the distribution shown in Figure 4 is slightly skewed, it seems reasonable to expect that the most likely result for a new investment will be about four units below the original profitability estimate and that about one-third of the new business projects entered into by a company could be expected to fall six to seven units above or below this value. From spot checks of other projects it appears that the standard deviations shown in Table 1 may be very general in usefulness and applicability. Although this review was based mainly on U.S. investments, similar or greater variability is expected in other countries.

Suppose a new U.S. marketing investment is estimated to show a 20 per cent rate of return, using estimates as realistic as possible. Using the data in Table 1 as a guide, the most likely actual result for this investment would be a 16.4 per cent return (i.e., 20-3.6) or one chance in three of actually yielding either under a 9.9 per cent return, (i.e., 16.4-6.5) or over a 22.9 per cent return (i.e., 16.4 and 6.5).

If a cost-saving project were under study in which marketing variables were unaffected then the profitability performance could be expected on the average to be close to the original estimate, with a standard deviation of only ± 3 units, as shown by the "all other" category in Table 1.

It is clear from these findings that the marketing variables can have a great impact on the profitability of new business ventures. Not only do small changes in these factors significantly affect rate of return, but the variables themselves are subject to greater uncertainty than most technical elements, because of the many marketing influences outside a company's direct control. Because of this it will probably be very difficult for a company to significantly reduce the uncertainty in future profitability forecasts.

In view of the uncertainty in even the most thorough profitability forecast, some form of risk analysis is clearly necessary. There are a number of different ways of measuring project risk--of varying sophistication and usefulness. Three will be discussed--project sensitivities, decision tree, and Monte Carlo analyses.

SENSITIVITY ANALYSIS

This is a common technique for identifying the important variables affecting a project appraisal. In its simplest form a sensitivity shows the change in a variable necessary to cause a given change in profitability--a one unit or 1 per cent change, for example. This simplified approach is often used in the early stages of a new project development, when the likely uncertainty in most variables is unknown.

Typical sensitivities for chemical industry projects (4) are shown in Table 2. This study covers three types of projects, varying mainly in investment intensity per dollar of sales. The information selected is for new business investments earning a 15 per cent rate of return.

It is clear from this table that relatively small changes in price and idle capacity (i.e., sales volume) can have a significant impact on project profitability. Note in particular that price is the most important variable, in terms of potential impact on project profitability, and idle capacity or volume is probably next. Incidently, the volume sensitivity data are for a single year; a small continuing volume loss would significantly affect profitability also.

When a project is sufficiently defined for a capital appropriation, the sensitivities should be based on the likely uncertainty in the major variables, and show the impact of this uncertainty on expected profitability. This type of approach is commonly used in economic studies and reports to management because of its relative simplicity and ease of understanding, and its helpfulness in appreciating project risk.

The hard part of such sensitivity analysis is the determination of the likely uncertainty in each variable. The appropriate experts must be consulted and a consensus obtained. Usually, these estimates are based on what are called subjective probabilities, as they are derived from individual opinions and past experience, and are not subject to repetitive testing and measuring under real world conditions. Another problem is to assure that "likely uncertainty" means the same to everyone. The exact value does not matter as long as everyone uses it. For example, a typical uncertainty might be defined as the high and low values a variable has one chance in ten of exceeding.

DECISION TREE AND MONTE CARLO ANALYSES

These are more sophisticated and complex methods of measuring project risk, and are generally developed only in studies of major investments. Again, the hardest part is the development of the subjective probability estimates. Often a sensitivity analysis is done first, to identify the most important or dominant variables before either of these more complex methods is used.

Briefly, decision tree analysis is an orderly method of structuring a problem. The schematic form of a decision tree is shown in Figure 5. The tree has nodes that are under the control of the decision maker (the square nodes) and nodes that are not under his full control (the circled or chance nodes). For example, in a commercial development program, successive square nodes should be bench scale research, pilot plant research, market development and commercial plant construction, and the circular nodes, possible results and their probability of occurrence.

In preparing a decision tree, the analysis must answer a number of questions. For example, what choices can be made now? What choices can or must be deferred? How can choices be made that are based on information learned along the way? What information can be gathered purposefully and what can be learned during the normal course of events without intentional intervention? What experiments can be performed? What are the likely technical marketing, and economic results?

After drawing the decision tree, the analyst must assign probabilities to the branches leaving the chance nodes. This is based on subjective estimates, usually by a number of different people. This is the hardest part because of the difficulty in obtaining reliable estimates of uncertainty, but except for a possible increased number of choices is no more difficult than determining likely uncertainties in the more common sensitivity analysis.

The next step is to establish the cost and benefit of each effect. This is often done by using net present values. The optimal strategy is then calculated, i.e., the strategy that maximizes expected values. This is fine as far as it goes, but usually it does not go far enough, since the implicit assumption of risk neutrality by management, often doesn't apply. Thus if the path with the highest expected value also had the greatest risk, it might well be turned down, and a less risky approach taken. The logic of this is sound: the high risk, high gain path could have devasting consequences if unsuccessful, where as the low risk, low gain path while not offering great gains, avoids disaster.

A Monte Carlo analysis (Table 3) replaces the probability of selected outcomes at the chance nodes of a decision tree by a probability distribution of all possible outcomes. There are four main steps in such an analysis:

1. Obtain probability values for significant factors,

2. Randomly select sets of these factors based on chance of occurrence,

3. Determine profitability for each combination,

4. Repeat many times to get probability distribution of profitability results.

Again, the hardest part is the development of the subjective probability estimates required in the first step. This type of study is usually done only at the time of final decision when inputs are better defined and a sophisticated risk analysis is more meaningful. The uncertainties in the costs and revenues of a project proposal are usually converted into a probability distribution of net present values for the total project, without any quantitative adjustments for management risk attitude. Because of the number of calculations required, this technique is almost always done on a computer.

Both decision tree and Monte Carlo analyses are useful tools in project studies but are less commonly used in presentations to management. Some of the reasons for not presenting projects in this manner include questions on the reliability of inputs, management unfamiliarity with the techniques, the problem of too much information, and the fact that risk attitudes are not considered. The use of these tools in decision-making has normally been based on maximizing present values without regard to the relative riskness of the project being compared. Unfortunately, the investments being analyzed by these techniques, because of their magnitude, are just where management is not risk neutral. Thus, these techniques will not be fully useable until they can be corrected for management's attitude to risk-something that is still not available in quantitative form. In the meantime, involvement of management in the earlier input portions of the analysis can help them to better appreciate the riskiness of specific projects, and acid in determining the amount of risk acceptable to them.

MANAGEMENT ATTITUDE TOWARD RISK

A company's future profitability depends on the quality and number of investment opportunities that are generated in the organization. All projects which are generated at lower levels must pass through a filter of management judgment. If the filter is based on overly cautious judgment, many potentially acceptable projects are strained out. If the judgments are consistently optimistic compared to real performance, corporate profitability will suffer. As might be expected, real life usually shows a wide variability in management risk attitudes in a given corporation. To assure more consistent decision making, an overall strategy toward risk taking in a corporation is needed.

The technique of management science known as decision analysis, is slowly emerging from theory to practice, although much further work is required.

It is the logical next step in analyzing risk. Features of the discipline are the treatment of uncertainty through subjective probability and of attitudes towards risk through utility theory.

The concept of subjective probabilities is well known and has already been briefly discussed. Utility theory, a method of risk assessment by quantifying management's attitude to risk, will be discussed in the remainder of this paper.

Utility theory applies to individuals as well as firms and government bodies. It is only a fancy name for everyday behavior. Although a common concept, it has only recently been used in formal risk analysis.

To illustrate the concept, suppose you had a rich but eccentric uncle who died and mentioned you in his will. Instead of a simple bequest, he gave you two choices:

1. One million dollars tax-free (his estate would cover any taxes you incurred).

2. A 50/50 chance of $10 million, tax-free, to be determined by a single flip of an honest coin. If you won, you would have $10 million. If you lost, you would have nothing.

Which choice would you take?

This question has been asked of several groups of students and almost everyone took the $1 million. Yet with a 50/50 chance of $10 million, the expected value of the second choice is $5 million or five times that of the first choice. Why choose the lower expected value? Basically, it is a question of risk aversion. The pain of not receiving a sure $1 million (if we loose the coin flip) is usually greater than the extra pleasure of the added $9 million (if we win the coin flip). Of course, there are exceptions to this attitude since it is subjective, and depends on the individual and his present circumstances. Both the chronic gambler (risk lover) and anyone who is already wealthy (so his circumstances would be little altered by losing one million dollars) would be likely to accept the second choice.

A conceptual utility function is shown in Figure 6. The abscissa is in dollars. The ordinate is an arbitrary index of pleasure and pain, or "utility." The decreasing slope is typical for investment situations and indicates a nonneutral risk attitude.

Risk neutrality would be represented by a straight line, and would be more likely to apply to small investments, or when comparing high gain alternatives for larger investments.

The "best fit" curve must be experimentally determined by appropriate questions of the decision makers. The utility scale is personal and subjective, and there is no reason to expect one person's utility function to agree with another's.

A similar concept can be applied to business entities. As might be expected from the "rich uncle" example, utility is not a property of individual projects but of overall wealth level, or relative immunity to "painful" loss. Also, the utility function on risk attitude of a corporation will not only vary from firm to firm, but will vary with time within a firm, depending on a corporation's recent and expected performance, and with the individual making up a firm's management. For example, the risk attitude of a strong president would dominate over other managers. On the other hand, if management by consensus is practiced, it could be represented by a significantly different utility function.

Because of these problems, the development of a generally accepted corporate utility function is very difficult. However once accomplished it could be very useful from a project standpoint, providing a quantitative method of discounting for risk.

THE CERTAINTY EQUIVALENT IN PROJECT EVALUATION

Another term for the utility (pleasure or pain) of a given monetary gain or loss is its certainty equivalent. To illustrate its potential usefulness, let us assume that a corporate utility function has been developed for and accepted by a company's management in order to provide a consistent policy towards risk-taking.

Suppose the company now has three large mutually exclusive investment choices. The expected values of these projects when discounted at the minimum acceptable return are:

Project	Expected Value, $MM
A	13
B	23
C	20

Ignoring risk, Project B would be accepted as it has the highest expected value. Unfortunately, management also feels it is the riskiest project.

Because of the magnitude and importance of the investments, a risk analysis using certainty equivalents is desired. This method requires two steps:

1. Measuring Risk
   This is done by developing a probability distribution of project profitability, using Monte Carlo analysis based on subjective probability estimates of the uncertainty in the major variables.

2. Interpretating Risk
   This is done by the certainty equivalent method which converts each probability distribution obtained in the first step to a single value which properly incorporates management's actual risk attitude.

The Monte Carlo results for Projects A, B, and C are shown in Figure 7, in terms of project present values. The total area under each curve represents a probability of 100 per cent, indicating that there is no chance of project performance falling outside the range of the curves. Present value distributions are used since rate of return distributions become meaningless in the negative range.

Much information can be drawn from these profitability distributions, but unless management's attitude towards risk is known, it is still difficult to reach a decision. For example:

Project A — Least risk. There is no chance of loosing more than $4 million present value and only a 2 per cent change of a negative present value. The expected value is $13 million and the maximum that can be received is $29 million.

Project B — Most risk. The project has a 27 per cent chance of a negative present value, but also has the greatest possible gain. The present values range from ($32) million to $83 million with an expected value of $23 million.

Project C — Intermediate risk. There is a 17 per cent chance of a negative present value. The expected values range from ($25) million to $58 million with an expected value of $20 million.

This information is a great improvement over a single present value based on "best guess" forecasts for each project. However, from a decision-making viewpoint there still are difficulties:

1. Not all management is familiar with probability distributions,

2. Even after fully understanding the meaning of the probability distributions, individuals are generally not able to adequately weigh all the information in the distributions,

3. The information tabulated does not indicate a clearly preferred alternate,

4. Risk attitudes still need to be considered.

A formalized assessment of risk attitude as in the certainly equivalent (CE) method can overcome these problems by making it possible to convert the probability distribution of a project's probability into a single number. The CE can be viewed as the risk-adjusted average value of the present value distribution. When the CE is greater than zero, the project is a good investment. Where several mutually exclusive alternates are being considered, the one with the highest CE should be accepted. Where several independent projects are being analyzed, all with CE's greater than zero should be accepted.

Figure 7, also, illustrates the use of the CE method for the three projects shown. The effect of risk discounting, i.e., the difference between the certainty equivalent and the expected value is shown for each project. The difference becomes increasingly significant as risk increases. These curves also show that failure to take advantage of all available information could give different project rankings from that obtained by the certainty equivalent method. The rankings by three criteria are:

Ranking Criterion	Project Ranking
1. Expected Value	B, C, A
2. Minimum Chance of Negative PV	A, C, B
3. CE Method	C, A, B

If the CE method is as good as it sounds, why isn't everyone using it? The answer is simple--as far as the writer knows, no firm has established a generally accepted corporated utility function so that certainty equivalents can be calculated--although experimental work in this field has been undertaken by a number of firms. It is hoped that work in this area continues, as the CE method would not only make risk analysis more understandable, but also improves project selection by giving proper weight to all information in the probability distribution.

CONCLUSIONS

Competition for high returns tends to drive all of us into high risk areas. If projects are selected by ranking methods which are blind to risk, then the firm tends to select a "diet" of high risk activities, and the business becomes highly risky. Conversely, fear of unknown or unquantified consequences often leads to decisions to not implement projects, thereby eliminating the possibility of many sound investments, thus reducing overall profitability. Business can be less risky and perceived risk can be reduced if methods of measuring and controlling risk are better understood and systematically utilized. Not the least benefit is that firms with a sound risk strategy, should be able to attract more capital, and at more favorable rates.

Several proven techniques for measuring risk exist for use by the analyst or management. However, more work needs to be done on analyzing management's risk attitudes, so that the tools for measuring risk can be better utilized. This is particularly important as historical data has demonstrated the significant uncertainty likely to remain in even the most thoroughly studied project. Most of this uncertainty comes from market and environmental factors beyond the control of the decision makers or the firm.

Literature Cited

(1) Park. W. R., "Post-Installation Appraisals--What the Literature Tells Us," Hydrocarbon Processing, March 1965, p. 111.

(2) Surtees, R. C., "Case Histories Reveal the Story," Hydrocarbon Processing, March 1965, pp. 116-18.

(3) Transactions, American Association of Cost Engineers, 1976, pp. 138-42.

(4) Twaddle, W. W., "The Power of Certain Important Factors in Economic Evaluation," 1962 AACE Transactions, p. M-1.

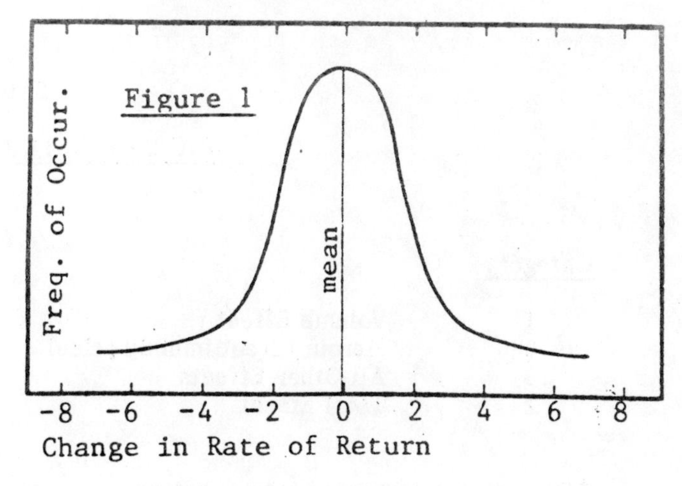

Figure 1. Effect of volume uncertainty on rate of return.

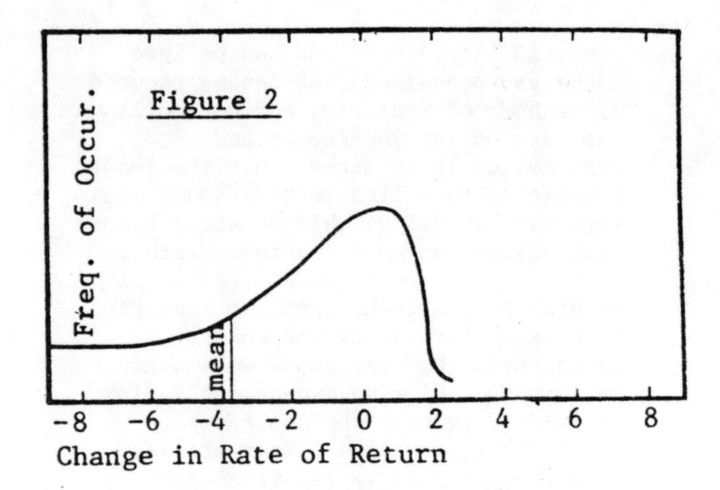

Figure 2. Effect of margin uncertainty on rate of return.

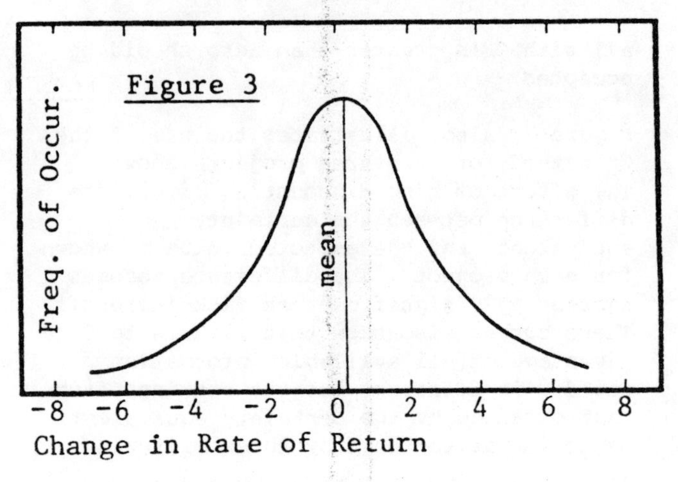

Figure 3. Effect of all other uncertainties on rate of return.

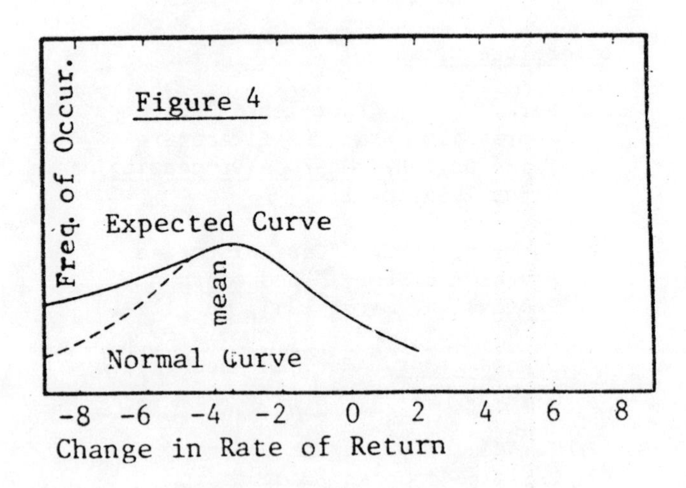

Figure 4. Total effect of uncertainty on base of return.

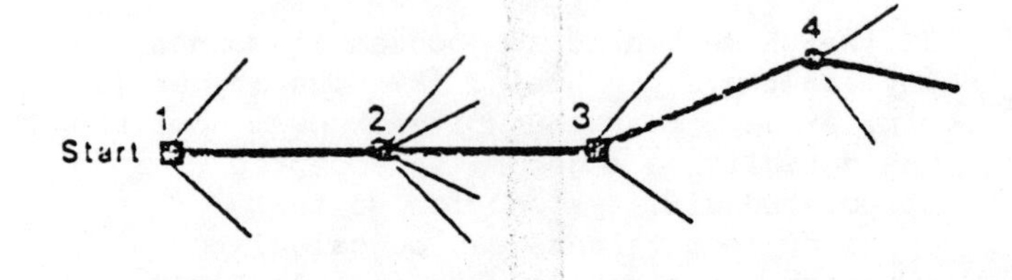

Nodes 1 and 3 are decision nodes
Nodes 2 and 4 are chance nodes

Figure 5. Schematic decision tree.

TABLE 1

RATE OF RETURN UNCERTAINTY

Category		Range* Minimum	Range* Maximum	Range* Mean	Standard** Deviation
1	Volume Effect	-10.0	+8.2	-0.1	± 4.0
2	Margin Effect (mainly price)	-15.6	+4.0	-3.8	± 4.8
3	All Other Effects	-5.9	+9.0	+0.2	± 3.0
4	Total Effect	-20.0	+15.3	-3.6	± 6.5

*Compared to expected rate of return.
**Assuming data are normally distributed.

TABLE 2

THE POWER OF IMPORTANT FACTORS

	Case A	Case B	Case C
Turnover Ratio, $ Sales/$ Fixed Capital	1.2	1.0	0.8
Required Variability per 1% Change in Rate of Return			
Sales Price, %	1.8	2.8	4.3
% Idle Capacity in 2nd Year	17.7	25.5	37.4
Fixed Capital, %	7.9	9.1	10.3
Startup Expense, % of Fixed Capital	12.9	15.1	17.1
Construction Period, Extension in Months	4.8	5.6	6.5
Startup Period, Extension in Months	3.8	4.5	5.1

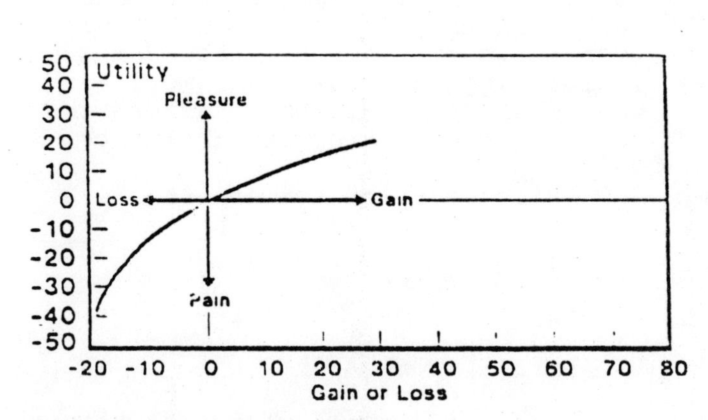

Figure 6. Conceptual utility function.

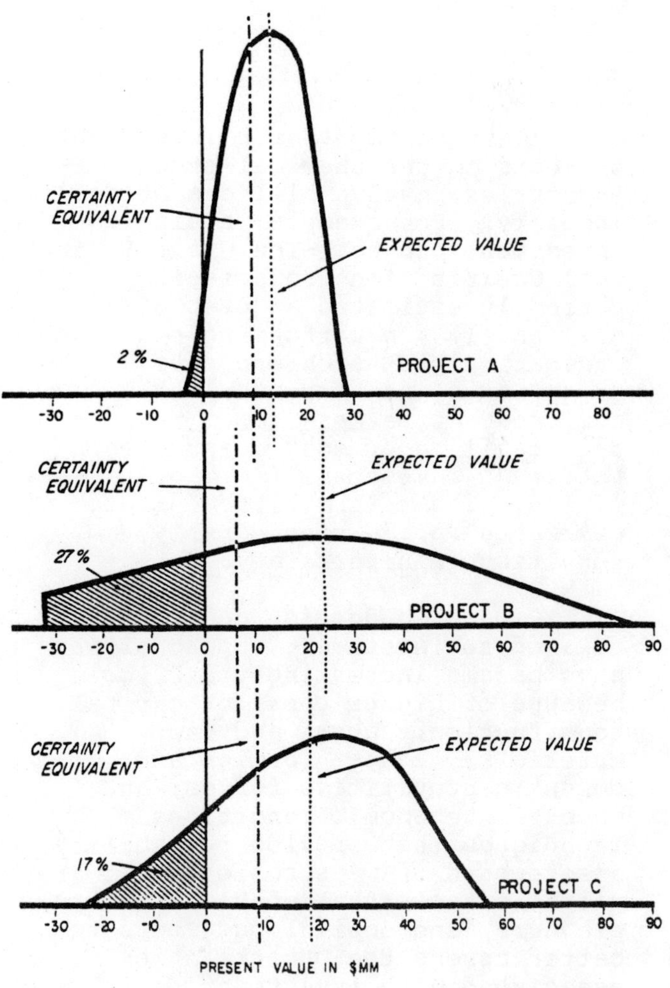

Figure 7. Probability distribution of present values for three hypothetical projects.

UNCERTAINTY ANALYSIS IN THE APPRAISAL OF CAPITAL INVESTMENT PROJECTS

The literature in several fields describes techniques for establishing uncertainties associated with economic feasibility estimates. Here the available methods and their underlying concepts are reviewed, and difficulties arising in their applications are identified. Determination of their suitability for specific purposes is delineated, calculation procedures are outlined, and suggested procedures for selection of a method are illustrated by examples. The paper is an overview of uncertainty analysis techniques and a plan for a users' guide.

VINCENT W. UHL

University of Virginia
Charlottesville, Virginia

STEPHEN T. SULLIVAN

System Planning Corporation
Arlington, Virginia

Capital requirements for investment projects in the chemical process industries involve billions of dollars annually. For example, a single investment project--The Great Plains Coal Gasification Project--is currently estimated at over $2.5 billion.(1) In performing economic evaluations for such projects a number of estimates are made for the components of a measure of merit--a profitability or cost measure such as return on investment (ROI) or payback period. The reliability of these estimates varies considerably, resulting in uncertainty.

Capital budgeting decisions related to investment opportunities have become increasingly difficult because of higher costs of capital, construction, labor, and raw materials; tighter foreign and domestic competitive forces; and uncertain economic conditions. Techniques that provide reliability assessments of measures of merit are valuable tools that enable managers, planners, and capital budgeters to better assess the uncertainties associated with investment decisions.

This discussion presents methods of uncertainty analysis used to assess the reliability of measures of merit. The conceptual approach underlying these methods involves analyzing the components of project benefits and costs and their interactions using systematic procedures. Each component influencing the measure of merit is varied in some fashion to assess the nature and degree of its influence.

First a brief summary of the methods in common use is presented. This is followed by a compendium of the field of uncertainty analysis used in capital investment appraisals.

ENGINEERING COST ANALYSIS

The profitability or cost measure is here referred to as the objective measure; it is represented by the objective function, and is affected by the objective function variables (e.g., sales price, operating expenses, and overhead). Estimates for these variables are used in the objective function to obtain values of the objective measure. Examples of commonly used objective measures are:

- Internal Rate of Return
- Return on Original Investment
- Average Rate of Return
- Payback Period
- Net Present Value or Present Worth

0065-8812-82-6201-0220-$2.00

- Profitability Index
- Uniform Annual Cost
- Levelized Cost

For grass roots project proposals, which are inherently more uncertain than modification or augmentation of existing projects, less rigorous indicators of merit--ones less analytically detailed or theoretically sound--such as payback period and return on original investment may be used. These do not include timing considerations and are often used to initially screen projects. Justification for the lack of rigor may be excused by a paucity of information, poor quality of data, or the requirement that grass roots proposals offer substantial profit margins on initial screening before a more rigorous, formal analysis is undertaken. For existing product lines, where data are often available and greater accuracy is desired, discounting techniques such as internal rate of return or present worth which incorporate timing consierations may be preferred.

In selecting a measure of merit, conceptual rigor and fastidious depiction of expected results are sometimes not the most vital considerations. Instead, the measure used may depend on its simplicity and the applicability of an evaluation methods consistent with the quality of input data. As in the computer maxim, "garbage in, garbage out," it generally makes no sense to use elegant evaluation methods on poor data.

AN OVERVIEW OF UNCERTAINTY ANALYSIS

Uncertainty analysis refers to quantitiate methods in which the components of a profitability or cost measure are examined to assess the likelihood and effects of obtaining results other than those expected from best-guess estimates. Uncertainty in the objective measure is the result of uncertainty in one or more of the variables in the objective function. Methods of uncertainty analysis investigate the behavior of those variables and their effect on the objective measure by decomposing its components, a process know as disaggregation.

There are a number of methods available for assessing the risk associated with a project. They range from rough indicators, such as payback period, to complex analytical methods utilizing large amounts of data and computer time, such as the Monte Carlo simulation. This discussion includes these methods in common use:

- Payback Period
- Risk-Adjusted Discount Rate
- Sensitivity Analysis
- Analytical Approaches Using Probability Calculus
- Monte Carlo Simulation

Payback period uses the time requird to recoup the original investment as a measure of the uncertainty in realizing return goals. The longer a project takes to recoup the original investment, the riskier it is assumed to be. This method requires relatively little data and is simple to apply. It assumes all estimates are correct, makes no provision for the time-value of money, and gives no measure of ultimate profitability. Inherently it provides no measure of risk.

The risk-adjusted discount rate method accounts for uncertainty by requiring that the project return a premium on invested capital to offset risks. It simply involves increasing the minimum acceptable rate of return (MARR) by a premium to account for risk. If the project is still profitable, the risk appears justified. This method is simple and provides the decisionmaker with an assessment of the safety margin the project offers.

Sensitivity analysis is a technique in which each objective function variable is tested at a number of values over its expected range while the other variables are fixed at their best-guess estimates to determine the effect of the range of a specific variable on the objective measure.

If the objective function has variables which are "random" (i.e., are described by probability density functions), the objective measure will be a random variable. Probability calculus can be used to analytically calculate the objective measure distribution from the objective function and the variable distribution functions. These methods are referred to as analytical approaches. It is generally necessary to assume that all variables are normally distributed to get manageable expressions for the results, as will be discussed.

The objective function is a simulation of the objective measure, describing cash flows associated with a project. If the variables in the objective function are random variables and if there is no straightforward way to analytically approach the problem, a sample value can be randomly taken for each variable from its probability distribution, and the objective measure can be calculated using these sample values. If this procedure is repeated many times, a distribution is obtained for the objective measure which theoretically approaches the analytical solution. This method is called Monte Carlo simulation.

CONCEPTS UNDERLYING UNCERTAINTY ANALYSIS

Uncertainty analyses are performed to identify and quantitatively describe the behavior of variables in the objective function that cause the objective measure to be uncertain. Since the objective function is rarely the most fundamental expression of the objective measure, variables that appear to be uncertain can be analyzed in greater detail by introducing the variables and relationships on which they depend. This substitution technique, referred to as disaggregation, provides the analyst with a means to focus on the specific components that drive the uncertainty in the objective measure. Figure 1 illustrates the additional insight into the underlying structure of the objective measure attainable through disaggregation.

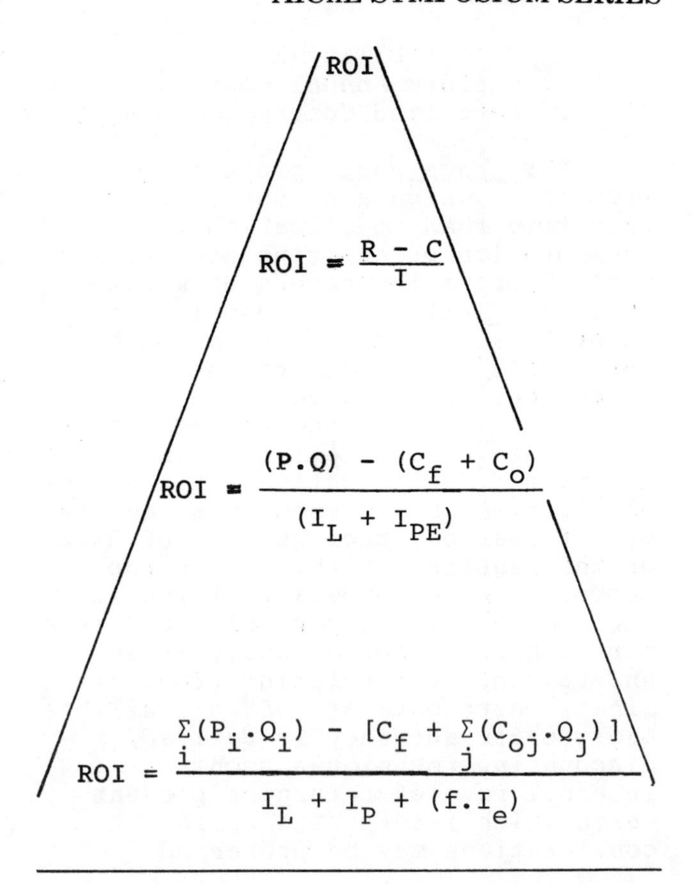

Figure 1. Disagregation of Return on Original Investment (ROI)

The objective measure, return on original investment, is at the highest level of aggregation, and can be disaggregated to sales revenue (R), expenses (C), and original investment (I). At a greater level of disaggregation, sales revenue is replaced by price times quantity (P . Q), expenses are broken down into fixed and operating expenses (C_f + C_o), and investment becomes outlays for land plus depreciable investment in plant and equipment (I_L + I_{PE}). At this point the relative impact of each of these six variables on the objective measure can be analyzed and assessed. The highest level of disaggregation in Figure 1 accommodates price differentiation, ties operating expenses to units produced and associated unit costs, and distinguishes indirect plant investment from outlays for equipment.

Dominant variables may be defined as those which may significantly affect the objective measure by resulting in values other than their best-guess estimates. If variables can be arbitrarily categorized into two groups--those which are dominant and those which are not--only the first group requires sensitivity testing. A variable may be dominant for two fundamental reasons. First, the objective measure may be sensitive to changes in a variable, such that a small change in the variable will have a substantial impact on it. An example is the interest rate used to discount project cash flows. Secondly, a variable is also dominant when it may vary over a wide range with little predictability, and significantly affect the objective measure as a result. An example of this is the estimate of sales volume.

Procedures for changing the objective function variable estimates may be categorized as deterministic or probabilistic. Deterministic methods assign test values to variables within some feasible range, but without regard for the likelihood of the respective values. Probabilistic methods incorporate probabilistic information about each of the variables, which yields a probability distribution for the objective measure. Sensitivity analysis is a deterministic procedure while Monte Carlo simulation is a probabilistic procedure. Deterministic analysis is often used to identify dominant variables variables so that they may be subjected to additional (probabilistic) analysis.

CORRELATIONS

Uncertainty analysis involves investigating the influence that each dominant variable has on the objective measure, and therefore requires changing the values of the variables in some fashion, depending on the technique involved. When variables are intedependent, a change in any correlated variable must be accompanied by commensurate chanegs in each of the correlated variables. For example, a 10 percent reduction in sales quantity would be expected to decrease production to some extent, though probably less than by 10 percent, thereby reducing total operating costs and increasing unit costs. The results of an uncertainty analysis which fails to identify and accommodate correlations among variables may invalid. This problem frequently proves to be the most difficult aspect in uncertainty analysis applications(2) Correlations may be involved for any variable that changes in value; they are often difficult to identify and quantify.

Relationships among variables are considered implicitly when best-guess estimates are made. For example, plant capacity will be based on a projection of expected sales, and will be primary determinant of the required investment. However, the relationship between some correlated variables may not be obvious. For example, when one variable is described by a probability distribution, the values of a correlated variable for corresponding points on the former's probability distribution may not obey a simple function such as a linear relationship. If expected sales volume is described by a continuous probability distribution, the corresponding variations in plant capacity and associated plant investment may not be obvious.

Identifying and quantifying correlations requires sound judgement and experience, and may be supported by collecting data on the variables to identify trends and quantify relationships. When a correlation is suspected, a rough test to determine if it could give misleading results is to use conservative (safe) estimates of the suspected variables.(3) If the project appears acceptable, these pessimistic values may determine that it can not be swayed by variable interactions. Similarly, projects deemed unacceptable can be confirmed as such by evaluation using optimistic values.

One way to avoid the necessity of identifying and accommodating correlations is to aggregate the objective function as much as is possible. Avoiding disaggregation limits the number of variables which may be correlated. However, this approach is not recommended in most cases, since it contradicts one of the purposes of an uncertainty analysis, i.e., to gain greater insight into the sources of uncertainty. Pouliquen states that "Limiting disaggregation can be considered only as an emergency measure in dealing with correlation. The advantage of risk analysis, after all, is that it permits disaggregation, and we want to retain this advantage."(4)

METHODS OF UNCERTAINTY ANALYSIS

Sensitivity analysis involves testing a noncorrelated variable, or a group of correlated variables, at a number of values over a reasonably expected range while fixing other variables at their best-guess estimates to observe their effect on the objective measure.

Sensitivity analyses are often undertaken initially to identify dominant variables. These variables can be disaggregated and a sensitivity analysis performed on the effects of the newly introduced variables. As the analysis progresses, this method identifies sensitive-variable combinations and gives quantified feedback on assumptions in the appraisal.

Sensitivity analysis is a flexible method for analyzing uncertainty. Any number of objective function variables can be tested, and in almost any desired combination subject to various correlation restrictions. A freqently used scheme is an interval in which the test variable is assigned a value at ± 5 percent, ± 10 percent, etc., of its best-guess estimate. Another method, based on combinations--setting all independent dominant variables except the test variable at pessimistic, best-guess, and optimistic values while varying the test variables as described above--provides a more exhaustive investigation of the behavior of the objective function.

As a simple illustration of sensitivity analysis, consider the objective measure, return on original investment, in Figure 2. The best-guess estimates are shown as R=100, C=85, and I=100. The figure shows the return on original investment when each of the three variables assumes values of 10 percent less and 10 percent greater than their respective best-guess estimates.

$$ROI = \frac{R - C}{I}$$

Best guess values: R=100, C=85, I=100

	ROI, in percent, for		
	0.9X Best guess	Best guess	1.1X Best guess
R, Revenue	5	15	25
C, Expenses	23	15	6
I, Investment	17	15	14

Figure 2. Sensitivity Analysis of Return on Original Investment (ROI).

As can be seen, the largest uncertainty due to a 10 percent change in value is from sales revenue (R). Expenses run a close second in impact, while investment variations produce relatively little change in ROI.

In many cases the objective function is a discrete series equation or an integral equation due to annual cash flow recognition and time discounting. Sensitivity analysis may also be used to test the response of these objective measures to trends, such as sales volume growth and decay that is faster or slower than anticipated. For example, the growth constant in the argument of an exponential function may be varied. These time-dependent test cases are dynamic in nature, as opposed to the static methods described above. For time-dependent functions, the mathematics may be rendered more tractable by utilizing Laplace transforms of the equations, which effectively reduce differential equations to algebraic equations. For finite lifetime projects a "truncated" Laplace transform is used, since ordinary Laplace transforms extend to infinity.(5)

The primary advantages of sensitivity analysis are its flexibility, versatility, and simplicity. It can accommodate almost any objective measure and is readily comprehensible. It does not require extensive quantitative background, and the results are generally in a form that makes interpretation a routine though often sizable task. Sensitivity analysis is easily adapted for computer applications; in fact, computers are necessary in all but cursory appraisals because of the large number of tedious calculations required.

The primary disadvantages of sensitivity analysis are the problem of correlations and the proliferation of data for all but relatively simple applications. Correlations will usually be important. When changing the value of an independent variable, any dependent variables affected by it should be modified accordingly, and vice versa, wherever they occur. As previously mentioned, failure to accommodate correlations can invalidate results. The output of a sensitivity analysis for a large or detailed appraisal can quickly become voluminous. Reducing this output to a form useful to decisionmakers can become a time-consuming and coslty task. To illustrate how the volume of output can proliferate consider the general method described above. Suppose that each variable is tested at three positions--a pessimistic value, the best-guess estimate, and an optimistic value. In n dominant variables are tested, the number of output values can be calculated by the statistical concept of combinations as follows.

$$\text{No. of output values} = \begin{bmatrix}\text{No. of test positions for each variable}\end{bmatrix}^{\text{No. of dominant variables}}$$

For example, if four dominant variables are tested, each at a pessimistic, a best-guess, and an optimiistic value, the result is

$$\text{Number of output values} = (3)^4 = 81$$

or 81 objective measure estimates. In a typical project, there will be more than four dominant variables, which confirms that data reduction may be a considerable task. In some cases, the more combinations of variables tested, the less clear the picture of the project becomes.(6)

Sensitivity analysis does not reflect the likelihood of obtaining any particular value for the objective measure. It does not incorporate probabilistic data and cannot create such information. This method merely illuminates the deterministic information embodied in the objective function and its variables. Sensitivity analysis can indicate which components to focus attention on, but cannot predict the likelihood of a given outcome. Though this is a disadvantage, the method is a powerful tool to determine the behavior of an objective function and may serve as a precursor to probabilistic methods for obtaining likelihood estimates.

PROBABILISTIC METHODS

Theoretically probabilistic methods involve replacing a variable with its probability distribution yielding a probabilistic expression for the objective measure. It is usually assumed that because a random variable behaves according to one of the standard probability distributions, it can be fitted to data by adjusting their parameter values and provide well-tabulated and well-behaved continuous analytical expressions.

Probabilistic methods are based on a branch of mathematical statistics, called the theory of expectations, in which certain characteristics of a random variable may be condensed into parameter values and used to represent these characteristics.(7) For example, the mean is a parameter representing the central tendency of a random variable, and the variance is a measure of scatter from the mean. The theorems of mathematical expectation allow parameter values (e.g., mean and variance) for a random variable such as the objective measure to be found from the parameters of the constituent variables. Therefore, if X is a function of Y and Z, the mean and variance of X can be found from the means and variances of Y and Z.

Theorems for addition and subtraction and approximate formulas for multiplication and division are given in Figure 3.

ADDITION

$$E(X+Y) = E(X) + E(Y)$$

$$\sigma^2_{x+y} = \sigma^2_x + \sigma^2_y + 2\sigma_{xy}$$

SUBTRACTION

$$E(X-Y) = E(X) = E(Y)$$

$$\sigma^2_{x-y} = \sigma^2_x + \sigma^2_y - 2\,\sigma_x\sigma_y$$

MULTIPLICATION

$$E(X\cdot Y) = E(X)\cdot E(Y) + \rho_{xy}\,\sigma_x\,\sigma_y$$

$$\sigma^2_{x\cdot y} = E(X)^2\,\sigma^2_y + E(Y)^2\,\sigma^2_x + 2\left[E(X)\right]\left[E(Y)\right]\rho\,\sigma_x\,\sigma_y$$

DIVISION

$$E(X/Y) = \frac{E(X)}{E(Y)}$$

$$\sigma^2_{x/y} = \frac{\sigma^2_x}{E(Y)^2} + \frac{-E(X)}{E(Y)^2}^{2}\sigma^2_y + \left[\frac{2}{E(Y)}\right]\left[-\frac{E(X)}{E(Y)^2}\right]\rho\,\sigma_x\,\sigma_y$$

Figure 3. Theorems of Mathematical Expectations

The analytical expressions for the random variables can be substituted directly into the objective function; this expanded objective function can be simplified, and an analytical expression for the objective measure will be the result. However, this is generally not a practical solution method since the complexity of the expanded objective function renders it unwieldy for analytical manipulation and the result usually does not conform to standard probability distributions--the properties of which are well known and generally amenable to manipulation.

To illustrate, consider the normal distribution, given by the equation

$$f(x) = \frac{1}{\sigma_x\sqrt{2\pi}}\exp\left[\frac{(X - \mu_x)^2}{2\quad\sigma^2}\right]$$

The product of two exponentials is another exponential; however, it cannot simply be assumed that the product of two normal distributions is another normal distribution. The normal distribution is a particular exponential function. By multiplying two normal distributions, f(x) and f(y) and grouping terms in the argument of the exponential, the result is often a normal distribution which may be represented by its parameters--the mean and variance--but it is not necessarily so.

As a second illustration of the complexity arising in analytical methods, consider the Beta distribution, given by the equation

$$f(x) = \frac{x^{\alpha-1}(1-x)^{\beta-1}}{B(\alpha,\beta)} \quad 0<x<1$$

where

$$B(m, n) = \int_0^1 u^{m-1}\,(1-u)^{n-1}\,du$$

It is much less obvious in this case that the product of two Beta distributed random variables is also Beta distributed.

THE ANALYTICAL APPROACH ASSUMING NORMALITY

The restrictions outlined above can be largely overcome if all random variables are, or may be assumed to be, normally distributed. The normal distribution is characteristic of random scatter around the mean value and has a number of mathematical properties which make it particularly amenable to analytical calculations. It is symmetric about the mean value, and the sum, product, etc., of normally distributed random variables is also normal in most cases. Table 1 illustrates some of the salient features of the normal distribution. Variable correlations can be accommodated by utilizing covariances, which describe the interactions of related variables.

Table 1
Normal Distribution

Percent	Range
99.0	$\pm 2.58\sigma$
94.45	$\pm 2.00\sigma$
95.0	$\pm 1.96\sigma$
90.0	$\pm 1.645\sigma$
80.0	$\pm 1.28\sigma$
68.27	$\pm 1.00\sigma$
50.0	$\pm 0.6845\sigma$

If all of the objective function variables as well as their products and quotients are assumed to be normally distributed, the theorems of mathematical expectation can be used to calculate a mean and variance for the objective measure, which is also assumed to be normal. The validity of these assumptions will depend on the skewness of the products and quotients. It is generally safe to assume that the mean of a function is approximately equal to the function of the means (the result of substituting the means for the variables) if the objective function is approximately linear.(7) In this case only a small degree of skewing, if any, results. If the objective function is highly nonlinear, the normal assumption of symmetry may not be justified. Correlations between functions add to the chance that the product will not be normally distributed. The expectation of the product of two correlated variables is not simply the product of their means.(8) For example, large numbers from each of the distributions are likely to be multiplied together when positively correlated, skewing the outcome to the larger numbers.

The skewness of a product of random variables is indicated by the coefficient of variation for the variables; defined as

$$\text{coefficient of variation} \equiv \frac{\sigma}{\mu}$$

The larger these coefficients are, the more skewed the product distribution will be, and the more the calculated variance using the approximate formulas will deviate from the exact variance. For cases where at least one of these coefficients is small, (generally less than 0.1 to 0.2), the calculated variance will be approximately correct.(9)

To apply the analytical approach assuming normality, variables are identified, and these random variables are assumed to be normally distributed. Best-guess estimates serve as mean values, and variances and covariances can be obtained from data or estimated.(10) If the product of two or more random variables is involved, the coefficients of variation should be calculated at this point to test the assumption of normality. If any are greater than 0.1 to 0.2, the error due to assuming normality may be significant and another method, usually Monte Carlo simulation, should be considered.

Once the means, variances, and covariances for the dominant variables are established, the theorems of mathematical expectation are used to calculate the objective measure in accordance with the objective function. When the mean and variance for the objective measure have been determined, the cumulative distribution can be calculated using a cumulative normal distribution table. These tables give the cumulative probability as a function of the variable X. They are usually in standarized form, which means that X has been made dimensionless by the transformation

$$Z = \frac{X - E(X)}{\sigma}$$

As an illustration of the analytical approach assuming normality, consider the objective measure, return on original investment, in which sales revenue (R) and expense (C) are considered dominant (Figure 4). The parameter values for each of the dominant variables are listed in the table, and are used in the formulas for expectation and variance to obtain parameter values for the objective measure.

$$ROI = \frac{R - C}{I} \qquad I = 100$$

	Mean (μ)	Variance (σ^2)
R	100	25
C	85	15

$$E(ROI) = \frac{100 - 85}{100} = 0.15 \text{ or } 15\%$$

$$Var(ROI) = \frac{25 + 15}{(100)^2} = 0.004 \text{ or } 0.4\%$$

Figure 4. Analytical Approach Assuming Normality

Analytical approaches based on mathematical expectation provide simple and useful probabilistic results. However, they require considerable familiarity with the mathematics and can be time consuming for all but simple applications unless the problem is simplified by assuming that all variables and results are normally distributed. In the latter case, the reliability of these techniques are determined by the validity of these assumptions. Results are generally acceptable if the objective function is approximately linear, and the coefficient of variation for at least one of two variables in a product or quotient is less than 0.1 to 0.2. If these criteria are not met, another method (probably Monte Carlo simulation) should be used.

MONTE CARLO SIMULATION

Monte Carlo simulation involves a sampling technique, almost always carried out on a computer, that randomly and iteratively selects sample values for each dominant variable from its respective probability distribution to obtain sample values for the objective measure. As the number of trials is increased, the resultant frequency distribution for the objective measure tends to converge to the analytical solution, thus simulating it by a numerical approach. The general procedure is outlined in Figure 5.

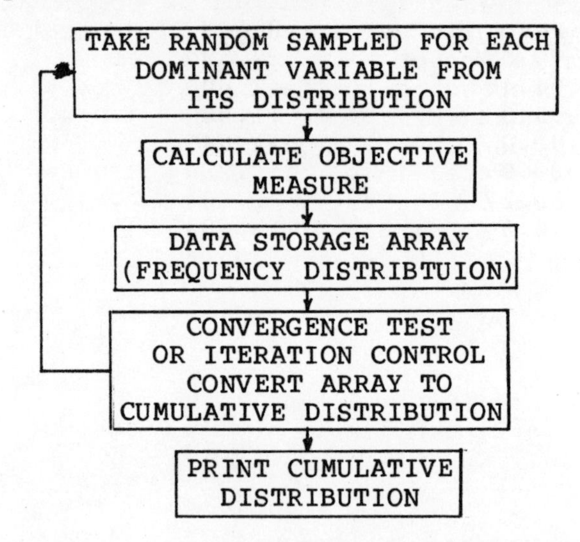

Figure 5. Monte Carlo Simulation - General Procedure for Sampling Technique

Monte Carlo analysis is a powerful technique for calculating probabilistic objective measures and can be applied to most uncertainty analysis problems. It is applicable with nonlinear and relatively complex functions, and they need not be of the same form, provided that convergence can be attained.

If an objective function has a small number of random variables and is relatively simple, a Monte Carlo analysis will be fast and straightforward, with the solution converging quickly. However, when many variables are involved and the objective function is not simple and well-behaved, Monte Carlo analysis can be complicated and expensive, with convergence being a major problem. Some distributions are particularly troublesome; e.g., those with zeros and those with long tails, and can make converquence difficult or impossible.

A number of methods for accomplishing the random sampling process have been proposed.(11) A basic version consists of applying a random number ranging from zero to one to the vertical axis of the cumulative distribution, which gives a corresponding value of the horizontal axis. When this process is repeated, the cumulative distribution generates sample values with frequencies corresponding to the density function for that variable. As an example, consider the random variable represented in Figure 6. A random number in the interval from 0.4 to 0.6 gives values of X between x_2 and x_3, and has a 20 percent probability of resulting. A random number in the interval from 0.8 to 1.0 has the same 20 percent probability of occurrence, but gives a much greater range of X values--from x_4 to u. There is a 60 percent probability that, if any one trial, the sample value of X is between x_1 and x_4. Sample values are randomly generated only for independent variables. Dependent variables are obtained from the independent variables using the quantitative relationships between them.

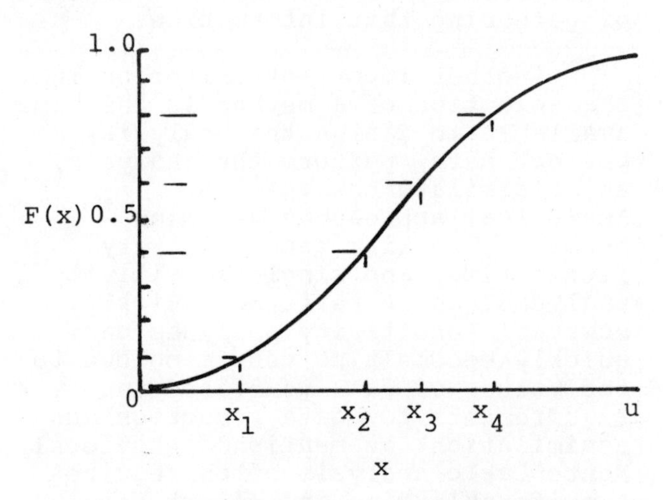

Figure 6. Cumulative Distribution

If normal distributions are used to describe the variables, a table of random normal deviates will provide random values for the equation

$$X = E(X) + \sigma_x Z$$

where Z ≡ random normal deviate

Standardized random variables are generated directly by random normal deviates according to a rearranged form of the above equation.

The number of trials required to approximate an analytical solution will vary with the particular objective function and variable distribution used, the goal being to obtain a significantly close approximation to the analytical result with as few trials as are necessary. The number of trials performed may be determined arbitrarily, or a test for convergence may be written into the procedure. Hess and Quigley found a significant improvement by using 100 trials as opposed to 20.(12) Other authors have used 500 to 1000 trials in their analyses. In general, results from less than 100 trials may be suspect, but convergence is often acceptable above the 100- to 200-trial range. A thousand trials seems necessary only in extreme cases, perhaps when an objective function is not well-behaved or when a large range of values must be considered.

Convergence tests are popular. and provide a practical method to determine the necessary number of trials for convergence.(13) A convergence test involves periodically calculating the parameters of the objective measure, and comparing them with those previously obtained. Generally, a trend or movement will occur initially, which will converge to a particular value. When the values of the parameter stop changing within given limits, convergence has occurred and the analysis is complete. If convergence has not resulted after a given upper limit of trials (e.g., 1000), the problem should be reexamined.

Correlations can be particulary troublesome in Monte Carlo analysis. Each trial theoretically represents a reasonably expected outcome composed of realistic combinations of the dominant variables. Only independent variables should be randomly generated, with correlated variables deriving from their relationships with the independent variables. Identifying and accurately representing these relationships, especially when there are a number of interrelationships among the variables, can be a complex task.

SOURCE OF INFORMATION ON THE UNCERTAINTY OF VARIABLES

Sources of information on dominant variables can come from a number of sources, including data from existing plant operations, market research information, and restrictions stemming from contractual agreements. The cost of obtaining additional information is balanced with the marginal benefit that such information is expected to provide. One particular method-- subjective probability analysis--deserves special mention due to its low marginal cost and high marginal benefit.

Contributors to a project appraisal and others having knowledge of or experience with the behavior of the components in an appraisal often have accumulated valuable information which may be utilized in an opinion analysis at low cost. In opinion analysis, a subjective probability distribution is constructed according to certain (often arbitrary, but consistent) procedures from the opinion of an expert or a committee of experts. These distributions are frequently needed in practice, and are an important component of uncertainty analysis.

Expert refers to a person who has experience in the field under consideration and, usually, with the particular problem at hand. This individual is often the most ideally suited source of information concerning the variable to the considered.

The mechanics of obtaining the distributions generally involves a series of surveys or questionnaires given sequentially to qualified experts. By repeating the estimation process and seeing their quantified opinions change over time, it is hoped that the views of the experts will be enhanced. To convert opinions into a subjective probability distributions, a distributional form is usually assumed, which is adapted to the subjective data. The assumed distribution may be either a density function or a cumulative distribution.

SELECTION OF A METHOD

The method of uncertainty analysis used will depend on the purpose and on the availability of additional information, time, and expense. The purpose of an analysis may be to exhaustively uncover potential uncertainties for a final design or simply to provide a preliminary estimate of the relative reliabilities of competing conceptual designs. Analytical approaches are often amenable to the latter case, while a Monte Carlo simulation would probably be required in the former.

Uncertainty analysis frequently requires additional data to determine ranges, probabilities, etc., for the objective function variables. Monte Carlo simulation generally has large data needs, and sensitivity analyses can also require large amounts of information. Consideration of various methods should include data requirements and the time and expense of gathering that information.

Another important criterion in the selection of a method is the time available to design the analysis, collect data, perform the analysis, and assimilate the results. Analytical approaches assuming normality require generally very little time, and single sensitivity analyses can be performed quickly. However, sensitivity analyses can quickly become time consuming due to the volume of data generated and requirements for data reduction and assimilation, as mentioned previously. Monte Carlo analysis often requires considerable time and effort to structure the problem, create the software, gather data, and assimilate results.

Closely allied with time requirements is expense. Sensitivity analyses and Monte Carlo similations are generally performed on computers, frequently placing substantial demands on them and requiring programmers and analysts. The expense associated with providing a team to identify pertinent factors, gather data establish variable distributions, create and test the software etc., can be significant. For these reasons, Monte

Carlo analyses tend to be particularly costly.

PRESENT UTILIZATION OF UNCERTAINTY ANALYSIS

In a survey by Petty, et al. (14), it was reported that payback period has been used most frequently as a measure of risk by the 109 firms surveyed. This survey rates the risk-adjusted discount rate as the second most popular method. About 30 percent of the respondents used simulation techniques, such as Monte Carlo and sensitivity analysis, fairly extensively, with over half of the respondents using them "in certain investment evaluations."

This industrial survey suggests that simplicity is an overriding concern, and that rigor is not always essential. However, as noted by the authors, ". . .almost half of the respondents expressed that [their] firm has moved to more quantitative and formal analysis." The Environmental Protection Agency, in analyzing alternative methods for accomplishing its goals, has expressed a need for more quantitative methods of reliability assessment in capital investment projects, which also indicates a trend toward these methods in case studies relating to both the private and regulated sectors of the economy.

In conclusion, perhaps the most significant reason that the more rigorous methods of project analysis are often avoided lies in the lack of a coherent, simple exposition of the means and methods of choosing and carrying out a particular analytical method. Part of this lack stems from the relatively new exploration of these areas; simplicity has not evolved. Part of the difficulty is simply the lack of a coherent, simple, widely accepted and understood vocabulary. Given the many different types of organizations using economic analysis and the many different arenas of application, this is not surprising. It is inhibiting, however. Part of the reluctance to use sophisticated analysis is that it, too, represents an investment which must be justified by a return. To date, the return has been difficult to quantify and thereby to motivate decision makers to commit themselves to its use.

ACKNOWLEDGEMENT

The authors wish to gratefully acknowledge the encouragement and support of the U.S. Environmental Protection Agency for this evaluation and exposition of the methods of uncertainity analysis.

NOTATION

C expenses including depreciation charges per year

I capital investment

P price per unit of product or service

Q annual rate of product or service

R revenue, or sales per year

Subscripts

e equipment

f fixed

j units produced

L land

o operating

P plant

PE plant and equipment

Common mathematical notation such as f(X), F(X), i, x, u are not specifically defined.

LITERATURE CITED

1. Interview with Ms. Jeanine Matte, Attorney, Office of the Assistant General Counsel for Procurement and Financial Incentives, Department of Energy, Washington, D.C.

2. Reutlinger, S., "Techniques for Project Appraisal Under Uncertainty," World Bank Staff Occasional Paper No. 10, (1970) p. 42. World Bank, Washington D.C.

3. Pouliquen, L., "Risk Analysis in Project Appraisal". World Bank Occasional Paper No. 11, (1970), p. 47. World Bank, Washington, D.C.

4. Ibid, p. 48.

5. Beenhakker, H., "The Engineering Economist, 20(2), 123 (1975).

6. Pouliquen, L., op.cit., p. 13.

7. Springer, C., et al., Probabilistic Models, Mathematics for Management Series, Vol. 4, p. 122, Richard D. Irwin Inc., Homewood, IL, (1968).

8. Goodman, L., Journal of the American Statistical Association, 55, 708 (1960).

9. Springer, C., et al., op. cit., p. 120.

10. Ferencz, P., Chem. Eng., 59(4), 44 (1952).

11. Hillier, F. and G. Lieberman, Operations Research, 2e., p. 625, Holden-Day, Inc., San Francisco, CA (1974).

12. Hess, S. and H. Quigley, Chem. Eng. Prog. Symposium Ser. No. 42, 59, 62 (1963).

13. Canada, J., Intermediate Economic Analysis for Management and Engineering, p. 283, Prentice-Hall, (1971).

14. Petty, J., D. Scott and M. Bird., The Engineering-Economist, 20(3), 167 (1975).

QUANTIFYING PROCESS PLANT VENTURE UNCERTAINTY

The uncertainty of satisfactory financial performance of a contemplated process plant venture is a critical factor in arranging for the capitalization of the venture. Project Finance and Uncertainty Analysis are discussed and quantified in a case study of an eastern oil shale plant using parametric studies, sensitivity analysis and risk analysis. The financial criterion used is the internal rate of return from life cycle costing.

R. G. WINKLEPLECK

Director Business Research
Davy McKee Corporation

A new venture involving a process plant project is in effect a separate business. The venture may be a step in the processing of a raw material or it may encompass overall processing to a final end product. The concepts of venture analysis are applicable for the contemplated project whether it is a cost center or a complete business entity.

This paper will focus on the financial aspects of a venture analysis. It is assumed that the venture sponsors have suitable assurances regarding technological and market considerations.

An essential ingredient in the check list of the investor is a sound financial analysis of the project. Quantification of the financial impact of changes in assumptions or expectations for the project as a business venture provide a measure of risk an investor assumes.

Uncertainty is associated with the design, construction and operation of the venture. Generally these are reflected in the capital and operating cost and in the ability of the plant to meet product output and quality expectations over some time period. The outcome of these uncertainties affect the financial outcome of the venture.

PROJECT FINANCE

In its purest sense, project financing (1) refers to the capitalization of an economic unit whose fixed assets are a specific project. The project is financed on its own merits. The creditors have no recourse to a higher business order. In practice virtually no project has been financed in the United States without some degree of surity for the creditors. Nevertheless this does not preclude the advisability or necessity for analyzing a prospective project as a stand alone venture.

The Economic Recovery Tax Act of 1981 includes significant increases in investment incentives through accelerated depreciation schedules, an extended tax carry forward period and increases in certain investment tax credits. Accelerated depreciation and tax credits have maximum effect early in the life of a project. However, the early years of a project are the most difficult financially. Costs are higher and output is lower during the start-up phase which might extend over several months. Concurrently, debt service payments will be higher under customary term loan schedules. A grace period for principal repayment offers relief during this pinch point.

The time value of money concept gives greater weight to financial events taking

0065-8812-82-6123-0220-$2.00

place sooner rather than later. Therefore, the tax benefits of high depreciation and tax credits are diminished if they are postponed. Profit before tax in any year which is insufficient to fully absorb the tax benefits leads to tax loss carry forwards.

When a project is analyzed as a stand alone business entity these deferrals are likely to occur. However, the tax act permits the transfer of depreciation and tax credit writeoffs to other entities permitting their immediate utilization. The same effect occurs in the project analysis when the pro forma financial statements of the project are consolidated with an economic unit with sufficient profit to fully absorb tax benefits generated by the project.

A project can require hundreds of millions of dollars capitalization. Such ventures generally comprise equity participation of sufficient financial stature to give at least tacit balance sheet strength to the debt portion of capitalization. It is not uncommon to have debt account for some 70 percent of project capitalization. Such heavy leverage of the equity position in a venture places ever greater demands for a thorough venture analysis.

LIFE CYCLE COSTING

A new venture will be subjected to financial scrutiny by investors and lenders of capital. A broad gauge of their reception can be gained by comparing expected project performance with their financial objectives.

In the case of lenders, their objective is reasonable assurance of being repaid with interest. A measure of this assurance is the debt service coverage ratio. This is the ratio of project cash flow to debt repayment obligations in any given year.

Investors may have several objectives for a project. Certain strategic business issues may outweigh financial goals, but generally a financial "hurdle rate" must be met to gain investor participation.

The usual quantification of hurdle rate is equating it to the cost of capital. Unfortunately this seemingly simple equation contains a subjective component for a project. Cost of capital is the weighted after-tax cost of debt and equity. Debt cost is readily determined. Equity cost for a publicly owned concern can be estimated from the price or value that the investing public places on equity - the market price of its traded capital stock.

In a project the cost of the equity capital becomes unique to that project. Its cost is a function of risk. High risk requires the potential for high rewards, but the subjective assessment of risk/reward therefore makes the hurdle rate an imprecise value.

Risk is the consequence of undesirable project performance. Uncertainty is simply a lack of knowledge of how the project will perform. (2) If we can quantify the uncertainties of a project and their financial ramifications, we have provided the investor a measure of his risk. The limits of risk for an investor usually go beyond the project investment to at least a tacit pro rata backing for the debt.

Having established the hurdle rate for a project, it remains to measure the project against that rate. An accepted approach to this is to use the basic concepts of life cycle costing. (3)

Life-cycle-cost in a venture analysis is the effective cost of producing a product over the life of the project to achieve a given internal rate of return on net funds into and out of the project. By definition the internal rate of return is that interest rate which discounts the future net funds to a net present value of zero. (4)

For simplicity, we can assume 100 percent equity financing and no inflation. This presents a base case from which we can later depart to adjust to a more realistic situation. The yearly project cash flow is tabulated starting with the time cash flow commences - the initiation of the project. The subsequent cash out-flows during construction and net cash in-flows during operations are thus tabulated. The interest rate which makes the cumulative discounted cash flow over the analysis period equal to zero is the internal rate of return (IRR). This rate is calculated on a trial and error basis.

This process can be repeated using various product prices. Higher prices obviously result in higher IRR's. By developing a curve of selling price vs IRR we can then find the price which meets our preselected hurdle rate for investment. That price is the life cycle cost under the assumed conditions.

UNCERTAINTY ANALYSIS

In analyzing the financial uncertainty of a project the factors affecting financial performance should be itemized. A useful starting point in the analysis is to rank the financial impact of each factor. Sensitivity Analysis is the best technique for this purpose.

Under life cycle costing we focus on IRR as a measure of financial performance. The sensitivity of a project to changes in variables affecting capital and operating costs can be measured by their impact on IRR.

A quantitative index of sensitivity can be developed by calculating the absolute percentage change in IRR for a unit percentage change in a variable.

The index of sensitivity identifies the most significant variables. The project appraiser can then direct his investigations or judgement to these key issues of uncertainty.

The uncertainty of the variables can be quantified by establishing an expected range of values. A higher order of judgement can be exercised by assigning a probability of occurrence. This can be conveniently expressed as high, low and most likely values. Statistical analysis of these values is facilitated by assigning confidence limits, say 95 percent, to the high - low range. With the most likely value the probability curve is established.

Given the likely range of variables, the IRR could be determined at the extremes of all variables. However, the likelyhood of simultaneous occurrence of extremes is remote. The resulting range of IRR would be quite large, and undoubtedly be disturbing to the appraiser while in fact bearing little relationship to the real venture dynamics.

A better way to examine the effect of variable ranges is to use a Monte Carlo (5) computer simulation model. In effect, the model calculates IRR with random selection of input data from the defined probability for each variable. The resulting probability curve of IRR then encompasses 95 percent probability because it is based on the same confidence limits for the input data. Consequently, the simulation model permits the development of a risk profile for the venture.

The risk profile constitutes a higher order of judgemental input for the project appraiser. Despite the sophisticated analysis it obviously can not dictate actual outcome. The root source of data is from forecasts.

The development of forecasts and expected ranges of values call on forecasting techniques. A rigorous analysis of faulty forecast data serves no useful purpose. Therefore it is incumbent on analysts to present a sound basis for the data offered.

VENTURE DYNAMICS

The application of uncertainty analysis should reveal the economic and financial characteristics of a venture. The information should do more than meet a go/no-go test. Rather, the dynamics of changing data should be apparent as in sensitivity analysis.

The organizer of venture capitalization will tailor the capital structure to cash flow patterns which occur with time. The tax shelter characteristics are of vital interest in structuring a venture. Effective long term rate of return on equity and capital employed are also of interest; so are the operating financial ratios over successive periods.

The accounting vehicle which vividly presents the time pattern of venture financial fortunes is the Statement of Changes in Financial Condition. This statement shows the yearly sources of funds for the venture and their uses. The sources of funds must equal the uses.

Uncertainty can also be addressed through this statement. The safety factor on cash available for loan repayment can be determined. The margin of safety required by lenders is proportional to the perceived likelihood of venture insolvency.

Financial pinch points are likely to occur early in the life of a project for the reasons previously suggested. The forecasted need for interim short term financing reveals project shortcomings. This might be avoided by a restructured capitalization.

Pinch points farther out in time in a constant dollar analysis will likely be covered by inflationary effects. However, a constant dollar analysis usually reveals trouble spots. A sensitivity test of solvency to operating adversity will reveal investor risk during the critical time intervals.

A venture might display favorable overall financial characteristics and still present untenable short term risk for a certain investor. This situation is cited to draw attention to risk as it is perceived by the risk taker. Different appraisers of a venture have differing thresholds for tolerating risk.

In the final analysis the investor in a prospective venture must make the investment decision. The value of venture analysis and the quantification of uncertainty lies in providing information for decision purposes. The dynamics of a venture might be the catalyst that crystalizes the investor's heretofore unstated limits for risk-taking.

CASE STUDY

Some of the concepts discussed in this paper are illustrated in a study prepared for Buffalo Trace Area Development District, Maysville, Kentucky. (6) Data from the report are used in conjunction with the provisions of the tax act of 1981 to provide a current example of venture analysis.

The contemplated grass roots project encompasses strip mining 30,000 tons per day of oil shale, handling, preparation, aboveground vertical kiln retorting and down stream processing. The products from the project boundary are crude shale oil, high Btu gas, light oil, sulfur and export electrical power.

The estimated capital requirements for the project as of late 1980 are:

Mine and Plant Capital Cost	$505 million
Interest During Construction	126
Working Capital	17
	$648 million
Deferred Mine Capital Cost	$ 126 million

During nominal operation, 330 days per year, the annual operating costs are:

Consumables	$23.6 million
Labor & Overhead	23.6
Taxes and Insurance	15.9
Gross Operating Cost	$ 63.1 million
By-Product Credits	52.5 *
Net Operating Cost	$ 10.6 million
* High Btu Gas $2.85/[illegible] ($3.00/million Btu)	$ 20.5 million
Light Oil $0.16/L ($0.60/gallon)	19.0
Sulfur $0.066/kg ($60/Ton)	2.5
Electrical Energy $0.0275/kwh	10.5
	$52.5 million

The project can be analyzed from the above basic data. Although details are not given here, further data is used to delineate the investment tax incentives available for investment placed in service after 1985 under the 1981 tax act.

The incentives apply through the Accelerated Cost Recovery System, investment tax credits, depletion allowance and expensing of mine development and start-up costs.

The maximum advantage of tax credits and write-offs is attained by using them at the earliest date. In the base case we will assume the venture is structured to shelter from tax income which is derived outside the venture. In effect, this means the venture enjoys a negative income tax in the years it encounters an operating loss. We will also perform the analysis in constant dollars, i.e. no inflation. As discussed earlier, the assumption of 100 percent equity financing and constant dollars provides a simplified foundation against which changes can be tested.

The results of the life-cycle-cost calculations for crude shale oil from the proposed venture is graphed in Figure 1. Continuous compounding and continuous cash flow is used in this case study.

If the venture must stand on its own and not shelter income from outside the defined venture boundary, deferral of investment incentives occurs. This results in an increase in life-cycle-cost, approximately 20 percent in this case.

The prospective investor can make a preliminary judgément from this curve. By applying his IRR hurdle rate for a constant dollar situation he can determine the required 1980 dollar selling price of crude shale oil to meet his objective. Then, the venture analysis resolves itself to assessing the likelihood that the selling price can be obtained.

The sensitivity of the project economics to change in variables is quantified by the index of sensitivity. The results of a computer analysis of the tax shelter venture are tabulated in decending order of economic impact.

Variable - V	Index - I
Investments	.104
Selling Price	.070
Production Volume	.069
By-Product Credit	.033
Labor	.017
Consumables	.015
Ad Valorem Tax and Insurance	.011

$$\text{Where } I = \frac{IRR}{\frac{V_2 - V_1}{V_1} \times 100}$$

And V_1 = Low value of a variable
V_2 = High value of a variable

The sensitivity analysis directs the appraiser of this project to pay close attention to the validity of forecasts for investment, plant output and by-product credits. Since selling price also has significant impact, and is generally set by the market place, a great deal must be known about market factors and likely future price pressures.

One parametric analysis of the venture economics is shown in Figure 2. Here the impact of change in value of the by-product credits is displayed.

The most likely values for the operating variables may be reexamined based on these results and another analysis prepared. Once the most likely values are established a quantification of their expected range will facilitate a Monte Carlo computer simulation of project risk.

For purposes of illustration, the following ranges are applied to the venture variables. The ranges encompass a 95 percent probability of occurrence.

Selling price	\$0.26/kg (\$40/bbl) ±10 percent
Production volume	rating +10 percent -5 percent
Investments	\$505 million +25 percent -15 percent
Working Capital	13 percent of sales
Consumables	±5 percent
Labor & Overhead	±5 percent
Ad Valorem Taxes and Insurance	+25 percent -15 percent
By-Product Credits	+5 percent -20 percent

Results of the risk analysis are shown in Figure 3. The probability distribution and cumulative probability of IRR performance for the project are plotted from the Monte Carlo simulation of the venture under the assumptions stated previously. The cumulative probability or risk curve indicates that there is a 90 percent chance the IRR will exceed 12 percent, a 60 percent chance it will exceed 13 percent and so forth. The mean IRR is 13 percent with a standard deviation of 1 percent. It is incumbent upon the investor to assess these odds against his objectives and his attitude toward risk. This analysis adds to the quantification of the uncertainties associated with the proposed project.

The final aspect of the project examined in this paper is the financial dynamics. For this illustrative case we assume capitalization based on 25 percent equity and 75 percent debt. This high leverage of equity demonstrates the cash flow for the investor's account after provisions for loan interest and principal repayment. Again the analysis is done in constant dollars to measure the basic dynamics of the project. Terms of debt are taken at 12 percent interest with 20 uniform debt service payments over a ten year period after plant start-up. Interest during construction is part of the capital structure.

Highlights of the analytical results from the sources and uses of funds over the period of debt repayment are presented first for a tax shelter situation. The price of crude shale oil is assumed to be $40 per barrel.

The debt-service coverage is:

Year of Debt Term	Debt Service/ Sources of Coverage
1	1.6 (good)
2	1.5
3	1.4
4	1.4
5	1.3 (borderline)
6	1.2
7	1.1
8	0.9
9	1.1
10	0.9

The effective rate of return on equity is 24 percent. No short-fall of funds is encountered after the project starts operation. Therefore, there is no requirement for additional short term borrowings or equity additions. An exception occurs in the 5th and 10th year of operation when major plant investment is expected for mine equipment replacement. This is funded from retained earnings in the calculation for return on equity. However, additional financing is the likely source when that time comes.

For comparison the project is analyzed on a stand-alone basis. The deferral of investment incentives via tax relief has a marked effect. Return on equity drops to 10 percent. Debt service coverage drops to 1.2 for each year of outstanding debt. The marginal oil price for financial solvency for the venture is $35/bbl.

It may be noted that the case study is analyzed with constant dollar cash flows and debt interest generally associated with an inflationary environment. This is a common dichotomy that has entered the managerial decision process. Perhaps it can be explained as an attempt to deal with "knowns." (7)

Many other conditions, scenarios and what-if questions must be analyzed for the investor because uncertainty is very pervasive. All variables, and the relationship between these variables must be at least minimally investigated, questioning performance under different conditions.

LITERATURE CITED

1. Peter K. Nevitt, "Project Financing," P. 1, 2, AMR International, Inc. (1975).

2. Canada, J. R., "Intermediate Economic Analysis for Management and Engineering," P. 190, 191, Prentice Hall (1971).

3. Schmidt, Blaine A., "Preparation of LCC Proposals and Contracts," Proceedings of the Annual Reliability Maintainability Symposium, Washington, D.C., January 23-25, 1979, IEEE P. 62-66 (1979).

4. Stermole, Franklin J., "Economic Evaluation and Investment Decision Methods," P. 187-189, Investment Evaluations Corporation (1974).

5. Hertz, D. B. Harvard Business Review 42, 95 (January - February 1964).

6. Davy McKee Corporation, The Cleveland-Cliffs Iron Company, Institute for Mining and Minerals Research "Synthetic Fuels From Eastern Oil Shale," Federal Grant No. DE-FC-4480R410185 (January 1981).

7. Crocker National Bank, "Inflationary Impacts on Project Evaluation; Real and Nominal Hurdle Rates," Project Financing Insight, 1, 1 (August 1981).

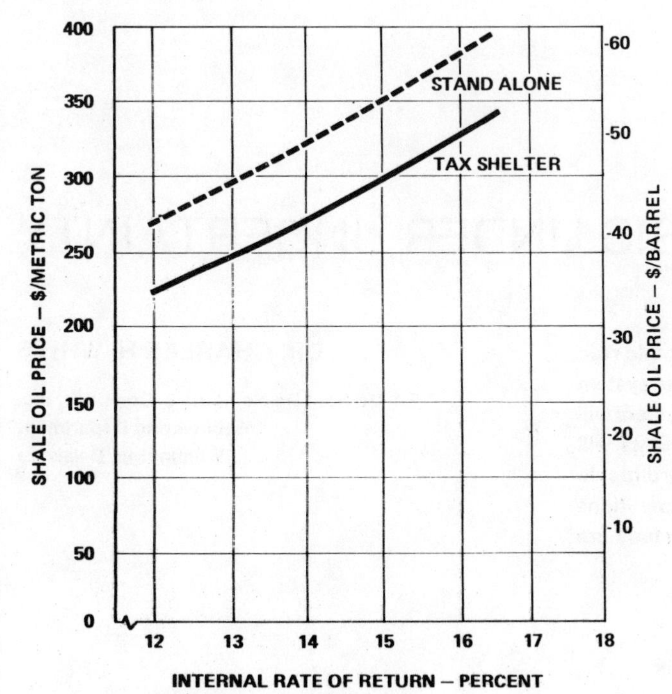

Figure 1. Life cycle cost 100% equity.

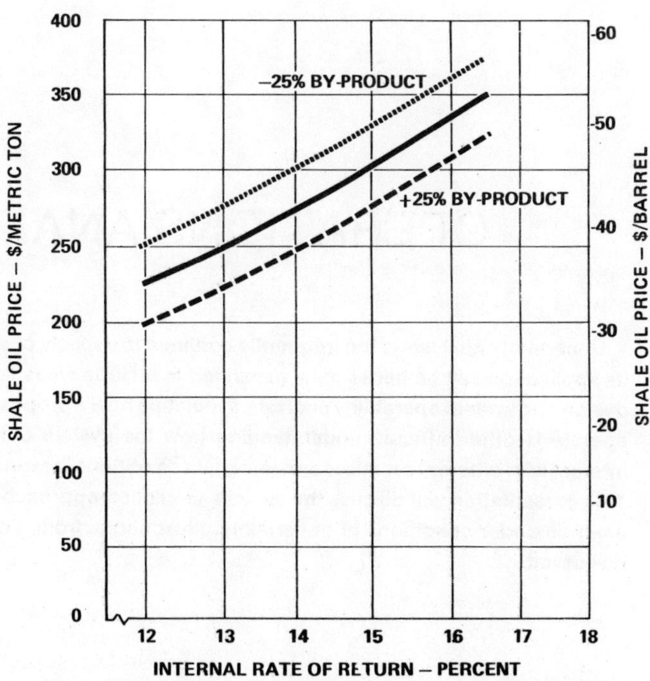

Figure 2. Life cycle cost with variation in by-product credit 100% equity-tax shelter case.

Figure 3.

```
        VENTURE ANALYSIS SYSTEM                                                      VENTURE IDENTIFICATION = BUFFALO TRACE

  INTERNAL VENTURE DCF/ROI RISK ANALYSIS PROGRAM-ABBREVIATED INPUT                   RISK ANALYSIS RESULTS
                                                                                     *********************

  NO OF      REQD FOR
  CASES      95PC CONF       MEAN       S.D.        ERROR
             (CASES)       (PC ROI)  (PC ROI)     (+ OR -)
   50          14             13.        1.0       0.2733

  THE MEAN DCF/ROI IS   13.PC WITH A STANDARD DEVIATION OF    1.0PC ROI.

  ROI        PROB          RISK
  (PC)     CURVE(PC)     CURVE(PC)
   11        10.00        100.00
   12        30.00         90.00
   13        42.00         60.00
   14        14.00         18.00
   15         4.00          4.00

             PROBABILITY & RISK CURVES

                                                                                                     1
             0    1    1    2    2    3    3    4    4    5    5    6    6    7    7    8    8    9    9    0
        0....5....0....5....0....5....0....5....0....5....0....5....0....5....0....5....0....5....0....5....0

  PC ROI = 10
```

OPERATIONS ANALYSIS UNDER UNCERTAINTY

Uncertainty analysis is too frequently confined to aspects of economic risk; its application can be beneficially expanded to include areas such as system design and system operability analysis. Predicting how a proposed system will operate is often difficult; understanding how the system components will interact with different premises and uncertain conditions is even more difficult. This presentation will discuss the system simulation approach to operations analysis under conditions of uncertainty. Illustrations from a case study are discussed.

DR. CHARLES H. WHITE

E.I. Du Pont De Nemours & Company, Inc.
Engineering Department
Wilmington, Delaware

Planning, designing, building, and operating production facilities are increasingly more difficult tasks in these days of large, complex, interactive systems. It is important to recognize and explicitly deal with several elements of risk in managing a commercial venture. Four major areas of risk can be identified: marketing, technical (process), investment, and system design and operability risk. These will be briefly described to set the stage for the topic of "operational analysis under uncertainty". It should be noted that in practice we do not often encounter and deal with these elements of risk in a nice, neat sequential fashion; instead, we must cycle forward and back as a venture evolves.

Marketing Risk involves our ability to sell the product in acceptable volume at an acceptable price. Early in the venture planning, we must evaluate the market demand for the product. Later in the planning cycle, we need to estimate the value-in-use of the product in different submarkets. In addition, we need to forecast the actions of competitors in these markets and consider problems such as competitive actions or early obsolescence of the product in the intended markets.

Technical Risk deals with fundamental process problems such as scale-up difficulties, low yield, short catalyst life, or poor product quality. It is useful to think of these problems as separate from those of equipment sizing or system throughput capability.

Investment Risk is difficult to separate from that of system design which is in turn impossible to separate from that of operational analysis. However, investment risk can be considered to include those economic uncertainties in the equipment investment and labor costs for a project assuming the system design is technically correct (risk free). For example, if we know the final design calls for a specific process facility with known equipment, there is still uncertainty in the final project economics due to variations in the investment, installation labor and/or project timing.

System Design and Operability Risk is the subject of the remainder of this paper. Predicting how well a proposed system will operate is often difficult. System operability is dependent upon many factors including:

1. Operability
   - component interactions
   - operating premises or philosophy
   - operating conditions
   - safety factors
2. Operational and maintenance policy
3. Recycle interactions
4. Cascading outages

To predict operability, it is necessary to understand how the system components will

0065-8812-82-6204-0220-$2.00

interact according to alternate premises and under different operating conditions. This is a complex task. There is also a tension between the somewhat contradictory needs to keep capital and risk low - not an easy task when the most obvious way to reduce risk is to increase capital outlay for "safety factors". Just when the driving force is to design and operate "minimum essential" systems, the complexity of the interactions and the elements of risk are pushing us toward putting in larger safety factors. The question is "Did we design a system that is likely to operate satisfactorily under the conditions we are likely to experience?" An alternative to capital safety factors is better predictive design.

Operational and maintenance problems should be considered in the design phase of any large processing or manufacturing facility. This is especially important when major recycle flows can cause interactions of the production facilities. For instance, a forced outage of one unit can cascade through an entire system affecting all of the units. One possible design response for reducing system interaction is to add additional spare equipment. Others are to add in-process buffer storage between units and to adjust rates in an attempt to keep operating units on-stream until the down unit is operable. Generally, these design responses for decreasing operability risk involve increases in capital outlay.

When considering operability from a design standpoint, some questions are: 1. what equipment should be installed, 2. what size storage buffers should be provided, and 3. what range of operating rates will provide reasonable flexibility? From an operations standpoint, some questions are: 1. under what circumstances should the unit operating rates be adjusted, 2. how often will there be a need to adjust the rates and by how much, and 3. how often will an outage of one unit cascade through part or all of the system?

The ultimate system question is: given the specific set of equipment, storage, maintenance and operating staff, and the specified control policies, what is the likelihood of the system meeting the planned output rates? An effective way of addressing this question in the presence of uncertainty is the subject of our discussion.

The specific concerns included in system design and operability risk are:

1. Do we need more equipment?

2. Do we need better controls or operating procedures?

3. Do we need more labor in the functional areas of operation, maintenance, materials handling, etc?

4. Do we have sufficient in-process storage to adequately buffer against problems such as equipment outages, rate mismatches, and demand fluctuations?

5. What is the impact if parameters such as rates, yields, or unit availability change?

The foregoing comments on system design and operability risk may be summarized as two paramount points:

1. System operability must be explicitly considered in the system design phase.

2. System design and operability risk interacts with investment risk.

If operability problems necessitate design changes, we should consider the impact on the required investment (in fact, we may need to make tradeoffs).

The method by which system operability is explicitly studied is by means of discrete system simulation, a modeling approach wherein a detailed procedural model of a system is developed and run. The objective of such a model is to generate a chronological state history of how a specified system will change through time starting from a specified initial state. Note that we are working with a procedural model which is different from a mathematical model. A mathematical model is a set of analytical or differential equations which describes a system and/or how it changes. A procedural model is a logical model which when executed will behave "just like" the real-world system. It can contain complex, nonlinear and conditional data such as "if the tank between two units is getting full, slow down the first unit and speed up the second unit". As we shall see, such models can contain very complex contingency plans and thus explicitly deal with interactions and interferences of components within the total system. Further, these models are developed in a modular fashion so we can pull out a module and replace it with one with expanded detail. These models have sufficient flexibility that if a system aspect can be adequately described in words, then it can be incorporated into a model.

As might be suspected, these models can become quite expensive. However, they are quite powerful tools in reducing capital cost, particularly considering that uncertainty can be incorporated into the model and "worked on" like any other system variable.

A key point is that the operability of the system should be investigated during the design phase. As we note in Figure 1, the cost of a design change escalates quickly once we leave the design phase; whereas, the potential for cost reduction opportunities quickly falls off.

Since modeling the system can reduce the system cost _and_ decrease operability risk, the model analysis should be done in the preliminary design phase (possibly even conceptual design) in which basic equipment sizes and control strategies are established.

When discussing simulation, it is useful to distinguish between _static_ or steady-state simulation, where there is no explicit treatment of the time element, and _dynamic_ simulation where the behavior of the system through time is a major factor. In the classical static or Monte Carlo case, there is no time element and the simulation process is based around the Monte Carlo sampling procedure. In a dynamic system simulation situation, we explicitly consider the time element. In a system simulation model we describe the system by describing the components in the system (examples would be the rates and availabilities of equipment and the size of tanks), the layout and material flows within the system, the operating rules and controls, and the staffing levels if appropriate. We then specify an initial state and let the model generate a chronological state history of the system. Since the model is a logical one, we can handle either deterministic or probabilistic elements.

System simulation modeling is similar to laboratory or pilot plant experimentation in that we build a model of the system, observe its behavior, and get a sample point. If we start from a different initial state, we might obtain different results or if we have probabilistic elements in our system, we can get different results due to sampling. So, we build a model and run it enough times so that we feel we can predict with a reasonable likelihood how well the real-world system will operate.

Table 1 displays the four fundamental types of models which may be used to describe a system. Engineering generally deals with deterministic analysis and focuses on steady-state or transient behavior. Classical risk analysis considers probability but focuses on the steady-state case; the Monte Carlo estimation of the variation in the total cost of a project is an example. Modern risk analysis deals explicitly wity both the time element and with probabilistic parameters. System simulation allows us to consider stochastic systems wherein the parameter distributions vary through time. A simple example of a stochastic system is the arrival of customers to the coffee shop for breakfast: not only are the arrivals random, but they all occur just when I want to eat. Furthermore, this also seems to be just when the service rate seems to slow down! (One might say stochastic is a fancy term for Murphy's law.)

Let us shift to a simple example based on a study which considered the _design and operability_ of a chemical processing system.

Figure 2 displays a simplified flowsheet showing three process units (I, II, III), three storage tanks (A, B, C), and ten material flows (a...j). Our key questions include 1. what size tanks A and B are required to adequately buffer process Units I and II from forced joint outages, 2. what maximum instantaneous rate should Unit II have considering that the rate of Unit I is already fixed by stream f, and 3. what is the best control policy for Unit II since it is possible to vary its instantaneous operating rate over a range?

Unit I primarily produces stream f; by-product stream b is either cycled through Unit II to recover raw materials and produce product j or passed through Unit III to produce product i which is less profitable. If tank A is full, stream b is diverted through tank C to Unit III; if tank A is empty, Unit II is forced to shut down for a specified time duration. If tank B is full, Unit II is forced down while if tank B is empty, Unit I runs exclusively from virgin make-up stream a since stream e will drop to zero flow.

The sources of uncertainty are introduced in two ways: first, both Units I and II are subject to equipment outages and second, there is a question of just how Unit II should be controlled. The decision variables to be considered are the size of tanks A and B, the maximum rate for Unit II, and the best control policy (there were several suggested) for Unit II.

To summarize our unit outage investigations, we determined that other similar units could be used as sources of reasonable availability information. We gathered six months of operating data working with plant operating and process group personnel. It was found that Unit I experienced outages of two types and Unit II of one type. We then analyzed the time between outages, and the length of these outages, and found that all time periods were good statistical fits to exponential distributions with different parameters. (We used Chi-squared goodness of fit testing procedures.)

The results of the investigations are summarized in Table 2. Specifically, Unit I operates on the average for 270 hours between failures of type I and 318 hours between failures of type II; these two types of outages on the average are 26 and 30 hours in duration. Unit II fails every 120 hours and is out of operation for 26 hours.

The control rules for Unit II set the operating rate based on the level of both tank A and tank B. Little is contributed by discussing the control rules in detail so I shall simply note they were rather complex and evolved as cases were run.

The system simulation model consisted of several separate but interacting modular subroutines, five of which are really "submodels" characterizing activity of an important system part. The model modules included:

- System initialization
- Unit I outages
- Unit II outages
- Unit I operation
- Unit II rate adjustment
- Unit II operation
- Report generation

As you can see from this list, the unit outage and operations were in separate modules. Once the model was complete, process simulation was initiated: we specified an initial state, generated unit outages as time evolved, adjusted the flows based on unit status and tank levels as time evolved, and kept track of a number of measures of interest.

Since there are sources of uncertainty in this system we needed to run replications to get a measure of variation about our sample results. In this illustration each run simulated one year of operation assuming a specific tank A storage capacity, a Unit II maximum rate, and annual hours of availability for Units I and II. Each case run randomly varied the actual times between failures and the durations of these outages according to the statistical models developed (Table 2). These random variations are the system uncertainty being tested in this simulation.

The results are shown in Table 3. The size of storage tank A (column 1) is in terms of hours flow from Unit I; for instance, the first run sized tank A to hold 24 hours flow from Unit I (stream b). The maximum instantaneous operating rate for Unit II (column 2) is in terms of percent of the Unit I flow rate of stream b; the first runs used 115 percent so Unit II had a steady-state catch-up capability of 15 percent. The unit availibilities in the next column are in terms of annual operating hours over a 8760 hour year; the first three groups of runs show a 79 percent availability, based on 6935 hours in the 8760 hour year, and the last group (Case IV) considers the result of increasing it to 87.5 percent, by increasing available hours up to 7665 hours per year. (Case IV is similar to Case I in rates and tank sizes.) Under the output heading, we see unit availability in column 4 and percent material lost in column 5. The availability shown in the "theoretical" row is based on the input values.

The results for Case I (Table 4) of running five replications were average unit outage percentages of 11.4 and 20.9 which compare to the anticipated theoretical values of 12.5 and 20.8 respectively. The percent material lost column represents the percent of stream b material that had to be diverted through Unit III due to tank A being full. As expected, it differs from run-to-run but averages 6.94 which from an economic point of view represented a significant cost penalty as it means we lost the revenue from stream j and required more virgin material from stream a.

Cases II-IV (Table 3) used different tank size and rate assumptions, and Case IV demonstrates a significant reduction in material lost. These flow cases illustrate that we have been able to explicitly deal with unvertainty while considering system design and operability questions. Many more cases were considered than those discussed here, and the results were utilized to change the preliminary system design and to evaluate alternative operating policies.

We have used system simulation models such as that above for many applications including: 1. to determine the value of additional equipment and tankage in existing production systems, 2. to evaluate the effect of more sophisticated control systems, 3. to help size operating and maintenance crews in existing and proposed plants, 4. to evaluate proposed productivity improvement programs, and 5. to analyze complex materials handling systems.

The general objectives of system simulation models in system studies are to:

- quantitatively understand total system behavior
  - interactions
  - interfaces
- predict overall (total system) operability
  - sensitivity
  - effectiveness of controls
  - impact of contingency rules
- evaluate design and operating alternatives
  - system improvements
  - cost reduction
  - guide design development
- anticipate problems
  - operating personnel interactions
  - develop contingency operating plans
  - consider reliability and maintainability
- explicitly (quantitatively) evaluate the impact of uncertainty on all the above.

In summary, the key messages of this paper are that 1. all of the above objectives should and can be handled in the presence of uncertainty and 2. the problems of operability and maintainability should be considered in the system design phase for any complex production system. As we have discussed, the opportunity for cost reduction potential is highest in the early stages of a project; at the same time the cost of system design changes are lowest. Modeling techniques discussed above can incorporate the time element and uncertainty for complex system simulations, allowing faster design decisions and tighter design. The payoff is better, safer, cheaper plants which are easier to operate and maintain.

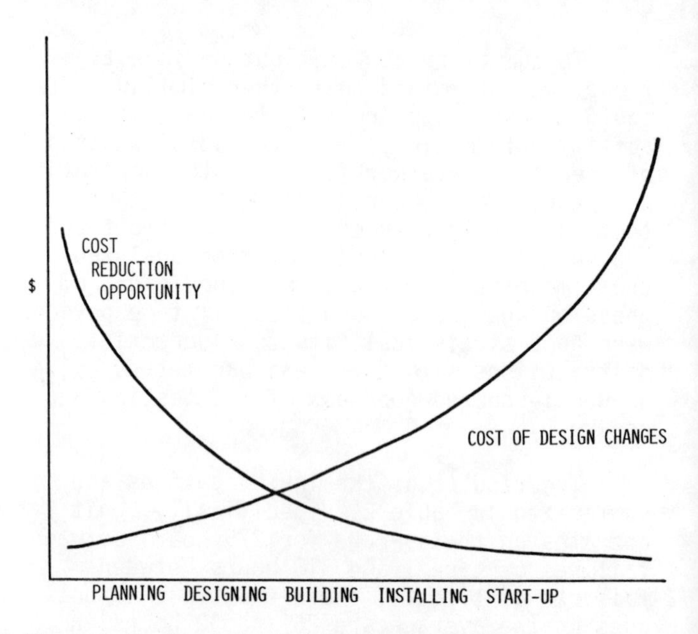

Figure 1. Cost aspects of project phases.

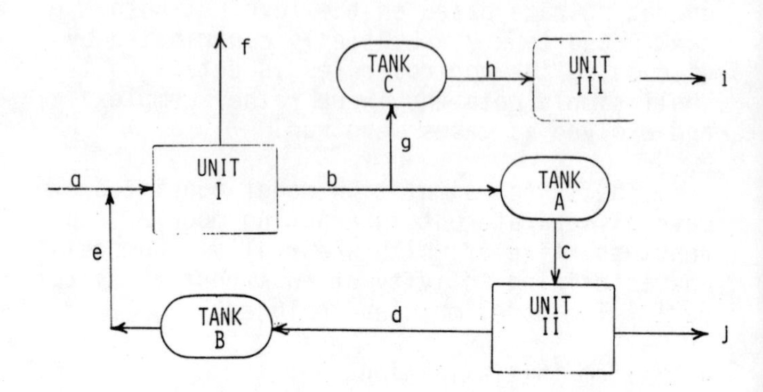

Figure 2. System flow sheet.

TABLE 1. MODEL ENVIRONMENTS

	STATIC	DYNAMIC
DETERMINISTIC	ENGINEERING "STEADY-STATE"	ENGINEERING "TRANSIENT"
PROBABILISTIC	CLASSICAL RISK ANALYSIS	MODERN RISK ANALYSIS (STOCHASTIC)

TABLE 2. OUTAGE ANALYSIS RESULTS

MEAN TIME BETWEEN OUTAGES (HOURS)	MEAN DURATION OF OUTAGES (HOURS)	
270	26	UNIT I, TYPE I
318	30	UNIT I, TYPE II
120	26	UNIT II, TYPE I

TABLE 3. SIMULATION MODEL RESULTS

INPUT				OUTPUT			
HOURS STORAGE TANK A	MAXIMUM RATE UNIT II	ANNUAL HOURS AVAILABILITY I	ANNUAL HOURS AVAILABILITY II	UNIT UNAVAILABILITY I	UNIT UNAVAILABILITY II	PERCENT MATERIAL LOST	
							Case I
24	115	7665	6935	.121	.213	7.1	Run #1
"	"	"	"	.139	.204	4.6	" 2
"	"	"	"	.096	.204	7.6	" 3
"	"	"	"	.116	.210	6.8	" 4
"	"	"	"	.097	.217	8.6	" 5
				.125	.208		Theoretical
				.114	.209	6.94	Average
				.018	.006	1.48	Std. Dev.
							Case II
12	115	7665	6935	.121	.213	11.6	Run #1
"	"	"	"	.139	.204	8.7	" 2
"	"	"	"	.096	.204	11.8	" 3
				.125	.208		Theoretical
				.119	.207	10.70	Average
				.022	.005	1.73	Std. Dev.
							Case III
24	125	7665	6935	.121	.213	4.8	Run #1
"	"	"	"	.139	.204	3.1	" 2
"	"	"	"	.096	.204	5.6	" 3
				.125	.208		Theoretical
				.119	.207	4.50	Average
				.022	.005	1.28	Std. Dev.
							Case IV
24	115	7665	7665	.122	.134	1.5	Run #1
"	"	"	"	.139	.130	2.3	" 2
"	"	"	"	.129	.117	1.4	" 3
				.125	.125		Theoretical
				.130	.127	1.73	Average
				.008	.009	.49	Std. Dev.

TABLE 4. CASE 1 RESULTS

CASE I

- TANK A HOLDS 24 HOURS.
- UNIT II MAXIMUM RATE 115%
- UNIT I AVAILABILITY .125
- UNIT II AVAILABILITY .208

RESULTS RUNNING GIVE "RUNS"

RUN NO.	AVAILABILITY I	AVAILABILITY II	PERCENT LOST
1	.121	.213	7.1
2	.139	.204	4.6
3	.096	.204	7.6
4	.116	.210	6.8
5	.097	.217	8.6
Average	.114	.209	6.94
Std. Dev.	.018	.006	1.48

UNCERTAINTY ANALYSIS OF A COMPLEX REACTOR SYSTEM USING MONTE CARLO PROCEDURES

RALPH H. KUMMLER
JOHN G. FRITH
CHEIN-SUNG LIANG

Wayne State University
and
Urban Science
Applications, Inc.
Detroit, Michigan

A major computer simulation was developed for the Detroit area's combined sewer overflow (CSO) program. The "system" which was modeled included over 1000 miles of six foot pipe in the sewer network, a treatment plant, an industrial, one dimensional plug flow river (Rouge River) and a major three dimensional river (Detroit River). The system simulation was run using a full year's rain fall as the driving force. The overall analysis of a given construction alternative required in excess of two hours on an Amdahl V-6. There were hundreds of input parameters with known and unknown uncertainties associated with each parameter. These models have permitted an ordering of the construction alternatives in terms of their benefit to the Detroit and Rouge Rivers.

Because there is an uncertainty associated with so many of the model assumptions and input parameters there is also an uncertainty associated with the model output used to rank the alternatives. This paper discusses the method developed to quantify the corresponding output uncertainty by examining the response of the output to variations of the input data. This sensitivity analysis includes (1) an absolute first order error analysis of the Plume Model of the Detroit River, (2) an intercomparison of independent models of the Rouge River, and (3) a Monte Carlo analysis of the overall set of models.

The results exhibit clear distinctions between most alternatives, but overlaps between some alternatives are also reported.

The benefit to the Detroit River of the various alternatives can be clearly defined since the uncertainty analysis shows that the cummulative frequency distributions for the different alternatives are separated well outside the error band. This result is found to be directly related to the fact that fecal coliform concentrations in the Detroit River are almost exclusively caused by CSO events.

The benefit to the Rouge River, however, is not well defined since the cummulative frequency distributions overlap considerably. This result is shown to be directly attributable to the uncertainty in the headwater concentrations which contain high levels of fecal coliform and oxygen deficits even in dry weather. Thus the benefit of CSO control on the Rouge is less certain.

The Detroit Water and Sewerage Department provides wastewater collection and treatment services over an area encompassing 650 square miles and including 3,200,000 people and over 1500 industrial dischargers. About 62 percent of the service area is served by separated storm and sewage sewers; the rest is served by combined sewers.

Once wastewater reaches the city, flow is generally towards the Detroit or Rouge Rivers where major interceptors divert dry weather flow to a single, large treatment plant near the confluence of the two rivers. Significant rainfall creates runoff in excess of interceptor capacity which, by design, overflows to the Detroit and Rouge Rivers at approximately 80 locations, called combined sewer overflows (CSO's).

The International Joint Commission (IJC) identified fecal coliform concentrations in the Detroit River as a problem that is related to CSO discharges (1). Under the Federal Water Pollution Control Act (PL92-500) Section 201 (2) Study just completed (3), this was confirmed. Similarily, the Michigan Department of Natural Resources (4) identified dissolved oxygen, fecal coliform, and total dissolved solids to be CSO related problems in the Rouge River.

In September, 1977 a Consent Judgement (5) mandated in item IV.B.2.a that the quantity and quality of combined sewer overflows from the City of Detroit be determined.

The objectives of this effort therefore became the following:

1) To determine the quantity and quality of combined sewer overflows to the Detroit and Rouge Rivers.

2) To quantify the impacts of combined sewer overflows on the Rouge River, Detroit Waste Water Treatment Plant, and Detroit River for the purposes of facilities planning, and

3) To provide a tool for evaluating the impacts of various CSO control alternatives on the Detroit and Rouge Rivers.

To comply with the Consent Judgement requirement to determine CSO "quality" and to have adequate information on the long and short term problems previously identified by others, the water quality parameters that were chosen include (1) BOD, (2) dissolved oxygen, (3) suspended solids, (4) dissolved solids, (5) volatile solids, (6) total phosphorus, (7) cadmium, and (8) fecal coliform.

The area modeled included the entire City of Detroit along with some small portions of a few suburban communities which totaled approximately 88,000 acres. Calibration data was collected by Giffels/Black & Veatch, the prime contractor, over a period in excess of twelve months (6).

0065-8812-82-6122-0220-$2.00

In the first stage of calibration, the model parameters were adjusted to predict as accurately as possible the total mass (flow and quality parameters) coming from the 80 CSO's and the treatment plant during the entire simulation period of nine months. Following this stage, the models were calibrated for total mass emissions for single events. The third stage of development/calibration focused on quantity and quality at specific overflows for all events.

MODEL DESCRIPTION

Executive Program

All of the models making up the Detroit Water Quality Information System are illustrated in block diagram form in Figure 1. The USEPA Stormwater Management Model (SWMM) RUNOFF and TRANSPORT blocks (7) are the commonly used models for predicting urban runoff and transporting flow to the rivers and the waste water treatment plant.

No major changes were required in the RUNOFF block for the planning level model. The TRANSPORT block had to be expanded to 700 elements to account for each overflow point. In addition, several flow routing options were added to account for the specific conditions in the Detroit system.

A dynamic model for the treatment plant (STPSIM II) was developed (8); the TREAT portion of SWMM was used for off site treatment analysis; the EPA models QUAL II (9) and the Raytheon model RECEIVE II (10), were used for the Rouge River; and a near shore plume model (11) and a finite difference model (12) for the Detroit River were developed.

As illustrated in Figure 1, the entire set of models interacts through an executive program which serves the following primary functions:

1) PRE-PROCESSOR: This module assembles time varying input data from the data base and verifies basic input data to the models.

2) POST-PROCESSOR: Model output is written to the data base and stored for future analysis. In addition, the post-processor has graphic and statistical analysis capability which is used to summarize results.

3) DATA BASE: All of the data required for the models is stored in a master data file that is accessed through indexed read and write statements. This enables efficient storage and retrieval of the model output for selective future analysis.

4) EXECUTIVE PROGRAM: The executive program enables unfamiliar users to utilize the models in total or selectively through a series of simple queries at a terminal.

In the next sections, we provide a brief description of each of the submodels. A detailed description of each of the models, the field data used for calibration, and the results have been published elsewhere (8, 11 to 26).

FIGURE 1
MODEL STRUCTURE AND OVERVIEW

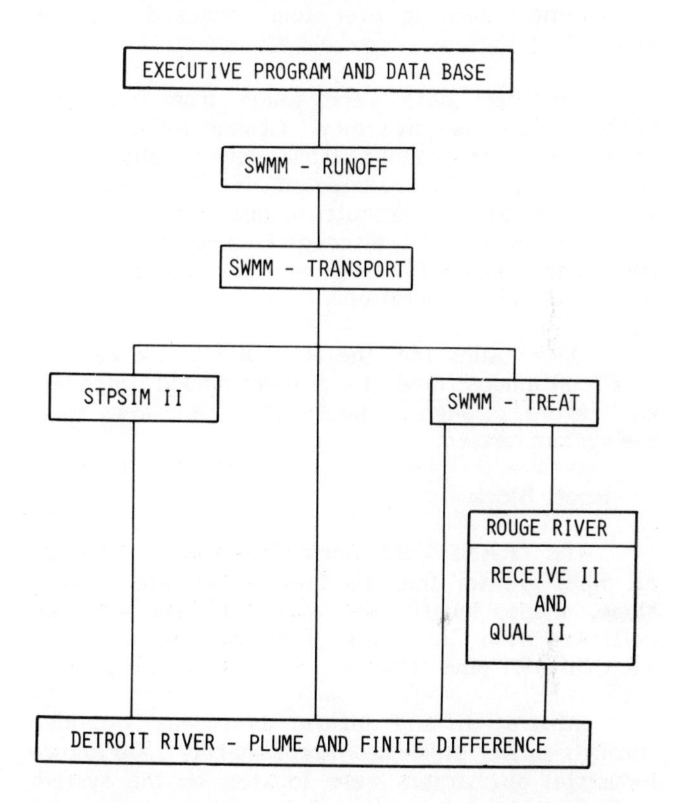

Runoff Block

The Runoff Model has been described by Anderson, et al (14). The function of this model is to describe the volumetric and species mass flow rates overland and through the minor piping system.

Most of the data for the planning level RUNOFF block was taken from existing sources. The City of Detroit was initially sub-divided into 39 sub-basins on the basis of sewer system flow and surface topography during Detroit's Segmented Facilities Plan (27). We were fortunate to have considerable rainfall data already collected through the joint efforts of the City of Detroit and the Southeastern Michigan Council of Governments

(SEMCOG) (28). The original limitation in the RUNOFF block of six rain gages was adhered to.

An analysis by Giffels/Black & Veatch indicated that the 1965 rainfall history was "typical" and was therefore selected for use in the alternatives evaluation. The actual rain history for 1979 and 1980 was used in calibration. In both cases, only the period from April 1 through December 31 was simulated, as snowmelt was not included in the Planning Level simulation. Thirty-two events occurred where rainfall was great enough to cause an overflow. Total rainfall during an event ranged from almost zero to over four inches during the year.

Land use data were taken from the 1975 SEMCOG land use inventory. Ground surface topography was determined from conventional topographic maps. The balance of the RUNOFF data was taken as the default values for the initial Planning Level model, although changes in the quality generation coefficients were eventually necessary to achieve calibration.

Once calibrated, the RUNOFF block was run in a "continuous" mode for a short period preceding each event through 24 hours after overflows from the system ceased.

Transport Block

The TRANSPORT block pipes were defined as all pipes greater than six feet in diameter. Size, slope, shape, length, and material data were all collected from City maps. The basic layout of the TRANSPORT pipe network is illustrated in Figure 2.

Infiltration and inflow data was provided through earlier studies. Approximately 1500 known industrial dischargers were located on the system and available data collected for use in the quality analysis. Considerable difficulty was encountered in obtaining reliable time varying data for the industrial users for nine parameters (including flow). In many cases, only average annual flow and annual mass loadings were available. Domestic dry weather flow was estimated using typical per capita contributions for quality and actual population figures from the U.S. Census.

Once the basic elements of dry weather flow were estimated Infiltration and Inflow (I/I), Industrial, and Domestic, actual plant operating records were used to proportionately adjust the estimated flows to match the actual daily flows observed prior to each event.

FIGURE 2
TRANSPORT NETWORK

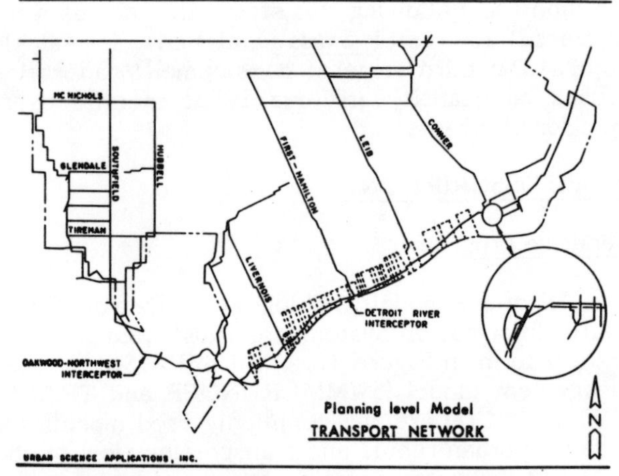

Figure 2. Transport network.

STPSIM II

An integral part of any storm water management system model is the treatment module. The size and complexity of the Detroit Wastewater Treatment Plant precludes the use of the existing SWMM treatment module. At the Detroit Waste Water Treatment Plant the accumulation of solid wastes in the liquid and solid processing streams is significant, and can also escape to the final effluent if the system is not strictly controlled.

A sewage treatment plant simulation model (STPSIM II) has been developed for this purpose as well as for the determination of plant sludge yield, chemical use and electricity consumption. The model dynamically simulates the behavior of unit operations in the solid and liquid handling trains of a conventional or oxygen activated sludge wastewater treatment plant. Provisions for predicting some process failures deterministically and simulating other process failures stochastically are built into the model. For example, studies have shown that turbulence in the final clarifiers of the activated sludge process induced by hydraulic loading changes can cause a process upset. This type of failure is modelled deterministically by STPSIM II. Failure of plant equipment, on the other hand, is random for a plant the size of Detroit's and can be simulated stochastically. In either case, process failure can be expected to significantly affect final effluent quality independent of influent quality for short time periods, such as during a storm event.

STPSIM II has been described and documented by Anderson (13).

Figure 3 is a schematic diagram for the processes simulated. Input to the model is supplied from the SWMM program. Hourly values for flow and quality are received by the treatment plant model as raw wastewater.

FIGURE 3
A SCHEMATIC OF THE DWSD PLANT LAYOUT AS MODELLED BY STPSIM II

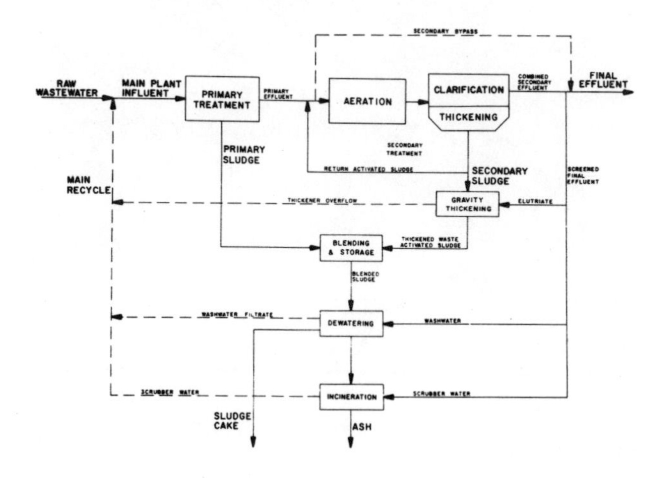

Rouge River Model

Two approaches were employed to model the effect of combined sewer overflow discharges to the narrow, industrial Rouge River illustrated in Figure 4. First, RECEIVE II, developed by Raytheon (10), was used to simulate the hydraulic behavior during wet weather. Water quality during this period was simulated in two ways. First, the quality block of RECEIVE II was employed, although the original numerical integrator was altered so that the program treated the 70 elements as a series of Continuous Stirred Tank Reactors (CSTR). Second, an analytic solution was developed which treated the quality calculation as if the river was a plug flow reactor. The former approach has been described by Harlow and Jamshidi, et al (19) and by Harlow and Frith (20) and the latter approach by El-Sharkawy and Jamshidi (17) and by El-Sharkawy (16). The two approaches were extensively cross checked for added assurance of accuracy. For the dry weather interim between storm events, QUAL II (9) was employed to calculate the steady state concentrations, a procedure necessitated by the benthic deposits which occur during the storm events and cause a dissolved oxygen sag long after the events terminate. A total of 1634 hours of wet weather data and 6600 hours of combined data was processed for each of the 70 elements.

FIGURE 4
A DIAGRAM OF THE ROUGE RIVER, ILLUSTRATING THE REACHES SELECTED FOR ANALYSIS TOGETHER WITH THE PRIMARY OVERFLOWS WHICH AFFECT EACH REACH

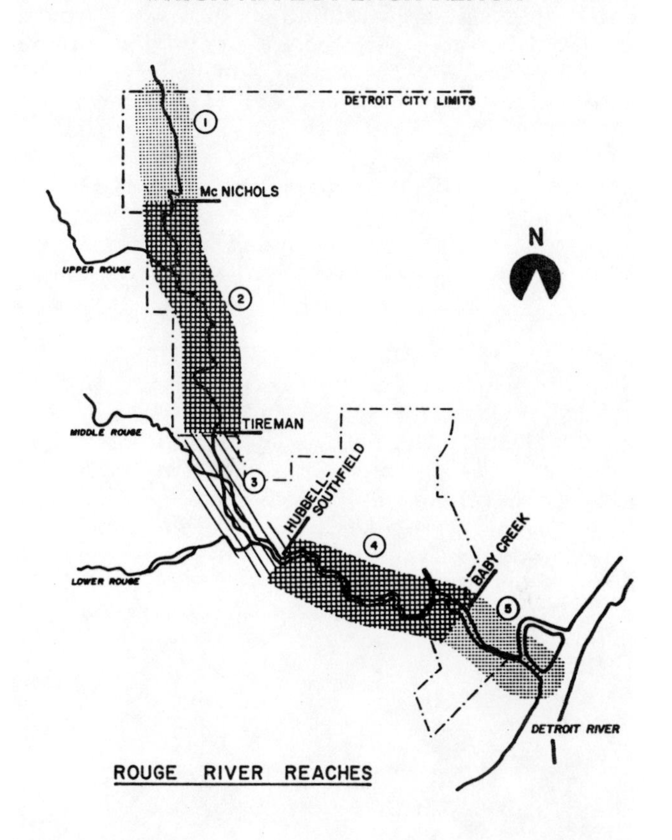

The Detroit River

Two different models were used to assess the effect of the combined sewer overflows, the treatment plant effluent, and the Rouge River on the Detroit River water quality. First, a finite difference model employed 74 segments to provide a planning level water quality analysis and to provide loadings to Lake Erie. This model was documented in Roginski and Kummler (25) and by Roginski (12). Model output can be obtained using computer plots such as those in Figure 5. The model is calibrated against the extensive EPA Storet data at the mouth of the river. Input is accepted on an hourly basis from TRANSPORT, RECEIVE II, and STPSIM II.

FIGURE 5
RESULTS OF THE FINITE DIFFERENCE MODEL FOR THE DETROIT RIVER

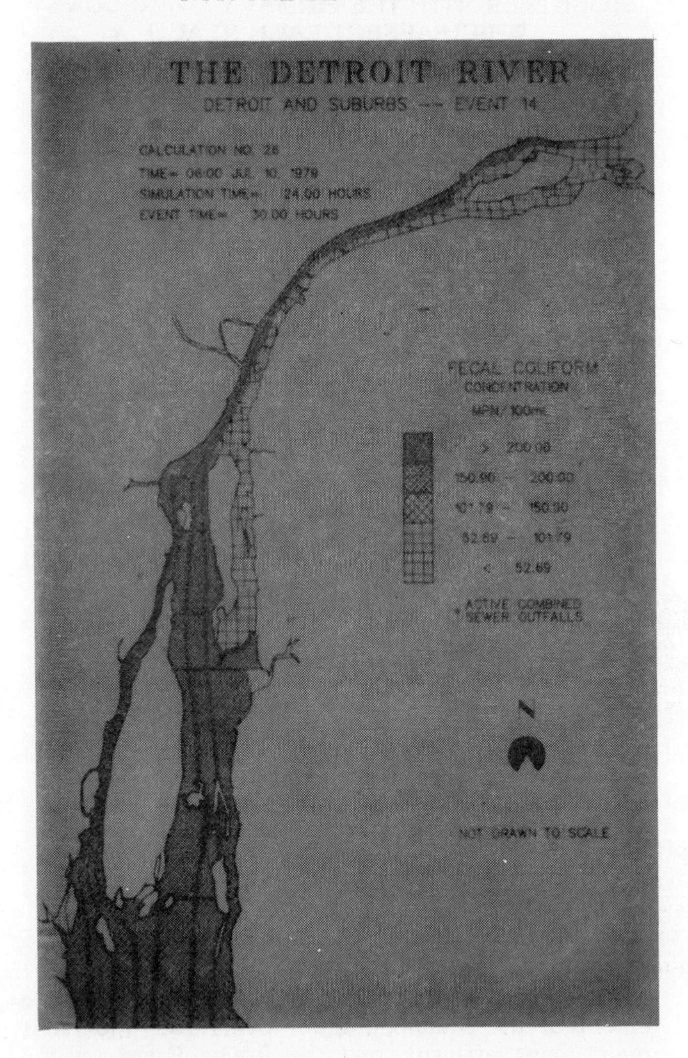

Second, to provide a more detailed analysis of the near shore concentrations where CSO's are expected to have the maximum impact, a Gaussian Plume Model was developed as discussed by Liang, et al (11) and by Liang (24). The plume model was calibrated using remote sensing multispectral scanner data of the DWSD plume taken by the Environmental Research Institute of Michigan. The region of the Detroit River treated by this model, which used 48 grid points covering about half of the USA side or one quarter of the river is illustrated in Figure 6. This grid was used to assess the magnitude of standard violations on the Detroit River and to perform the uncertainty analysis.

FIGURE 6
A DIAGRAM ILLUSTRATING THE DETROIT RIVER REACHES EMPLOYED IN THE ENVIRONMENTAL ANALYSIS ALONG WITH THE PRIMARY BASINS DRAINING INTO THE RIVER. REACH 1A, ABOVE CONNOR WAS NOT USED IN THIS ANALYSIS AS THERE WERE NO INPUTS IN THIS REGION.

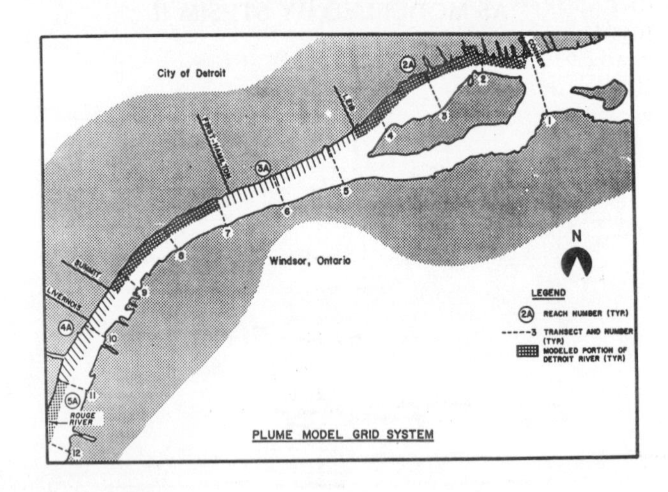

RESULTS

Alternative Analysis

The models described herein have been used to simulate water quality in the Rouge and Detroit Rivers for 26 different CSO control options varying from zero to 80% control of existing overflow volumes. Each alternative design was compared to the baseline "future no action" (FNA) alternative which includes some additional preplanned construction beyond the existing conditions. The projected benefits were then incorporated into a cost - benefit analysis (18) to determine the most useful alternative. The eight best alternatives had costs ranging from a low of \$142.4 million (capital) and \$68.2 million (annual) to a high of \$376.6 million (capital) and \$80.9 million (annual) (3). The overall results are given in Kummler (23). In terms of the "best" alternative, the results for the Detroit and Rouge Rivers for dissolved oxygen (D.O.), fecal coliform (F.C.), phosphorous (P) and suspended solids (SS) are given in Tables 1 and 2.

On the Detroit River the only significant problem is the level of fecal coliform counts. The best proposed construction alternative results in a six-fold improvement in the annual geometric mean concentration and a two-fold improvement in the number of hours of standard violation. On the Rouge River, an even bigger improvement is predicted for the fecal coliform annual mean concentration although only a 26% improvement is expected in the hours of violation. In both cases the standard is exceeded in the FNA alternative on an annual mean basis and is not exceeded under the best proposed alternative.

A comparison of the FNA alternative with the best alternative for several sections of the Detroit River in terms of the hours of violation is illustrated in Figure 7 (26). It is clear that marked improvements are predicted. For the Rouge River, the predicted improvements are considerably less as illustrated in Figure 8, but the annual geometric mean is greatly reduced as illustrated in Figure 9.

In both cases, considering the magnitude of the investment and operating costs involved, we must know the uncertainty of the predictions, so that we can consider whether the investment is justified.

TABLE 1

DETROIT RIVER					
	FUTURE NO ACTION (FNA)			ALTERNATIVE 12	
PARAMETER	STANDARD	AVERAGE CONCENTRATION	HOURS OF VIOLATION	AVERAGE CONCENTRATION	HOURS OF VIOLATION
D.O.-mg/l	7	9.75	0	9.77	0
F.C.-cts/100ml	200	240	529	42	252
P.-mg/l	0.003	0.003	177	0.002	134
S.S.-mg/l	25	11	32	10	20

TABLE 2

ROUGE RIVER					
	FUTURE NO ACTION (FNA)			ALTERNATIVE 12	
PARAMETER	STANDARD	AVERAGE CONCENTRATION	HOURS OF VIOLATION	AVERAGE CONCENTRATION	HOURS OF VIOLATION
D.O.-mg/l	5	6.34	1882	6.54	1558
F.C.-cts/100ml	1000	1137	2443	495	1808
P.-mg/l	0.12	0.44	6600	0.42	6583
S.S.-mg/l	80	26	24	24	12

FIGURE 7
ANNUAL HOURS OF FECAL COLIFORM VIOLATION IN THE DETROIT RIVER

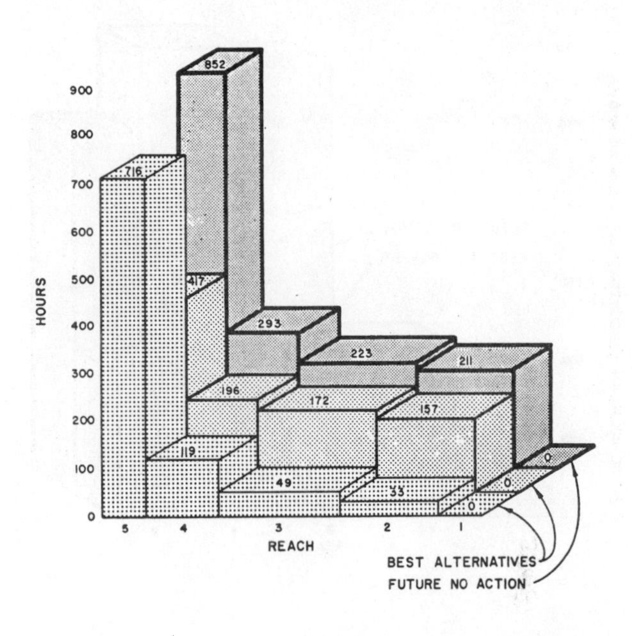

FIGURE 8
ANNUAL HOURS OF FECAL COLIFORM VIOLATION IN THE ROUGE RIVER

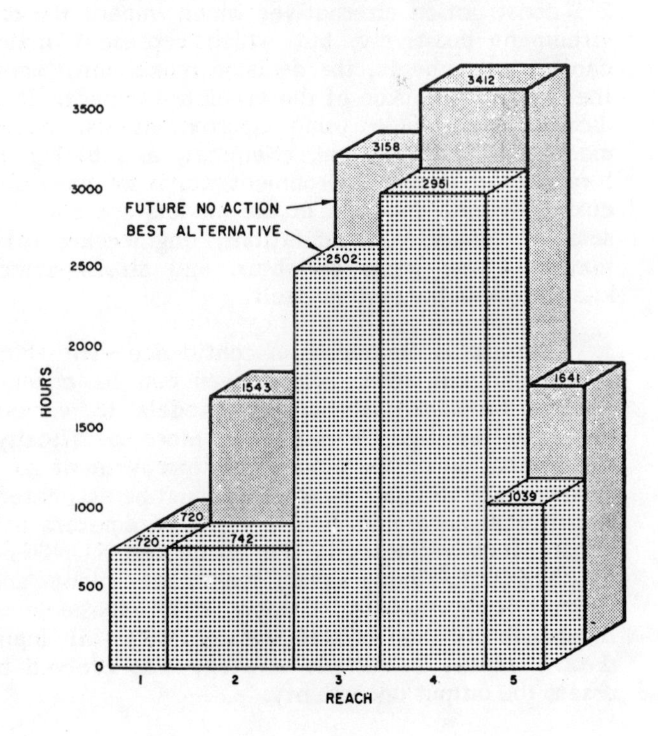

FIGURE 9
AVERAGE ANNUAL FECAL COLIFORM
IN THE ROUGE RIVER

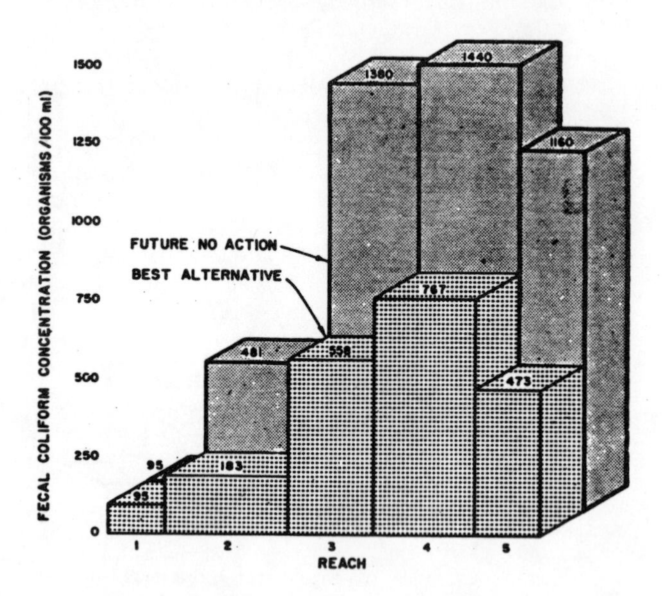

THE UNCERTAINTY ANALYSIS

To make critical decisions based upon Section 201 construction alternatives which impact the environment positively, but which represent major capital investments, the decision maker must know the relative precision of the predicted impact. Predictions are made using approximations, called models, of the physical, chemical, and biological interactions in the environment which we have discussed herein. Inherent in the models are complex sets of numerical calculations, engineering estimates of the input variables, and an imperfect knowledge of the system itself.

To assess the degree of confidence with which a screening of these alternatives can be accomp lished, the sensitivity of the models to various uncertainties must be measured. More specifically, the prediction of the amount of improvement as a function of the input uncertainty must be estimated. A full analysis would involve eight parameters and hundreds of variables in the connected RUNOFF-TRANSPORT, Rouge River, treatment plant, and Detroit River models. Clearly, it is impossible to independently and systematically vary all input data. Hence, a multiple strategy was evolved to assess the output uncertainty.

First, where possible, an analytic first order error analysis was performed. For the Detroit River, the important environmental impacts are described by the plume model which is an analytic description of the concentrations in the river due to dynamic CSO loading. For this case a full and rigorous analysis is possible and has been conducted. Such an analysis takes into account all sources of error, many of which do not pertain to the ranking of alternatives; hence, it provides an absolute error analysis as opposed to a relative error assessment.

Second, all models were coupled, and a preliminary sensitivity analysis involving the whole system using minimum and maximum bound techniques was conducted. The present work has developed from that effort, in which the upper bound method was clearly inadequate to assess the total probable error due to a large collection of input uncertainties.

The maximum bound approach greatly exaggerates the probable error. If the expected magnitude of a specific input variable is the value employed in the best estimate calculation, then choosing the maximum bound for all variables will give an incorrect picture of the probable error for the overall model. An alternative approach is the use of Monte Carlo simulation with stochastic model inputs. Explicitly treating parameter variability, this technique allows the combination of estimates of error from all sources in a finite series of computer runs. In this method, each error source is characterized by its frequency distribution rather than by a single value. The executive program chooses random values from the characteristic distribution for each variable. The model predictions are repeated using the randomly generated variable magnitudes and this results in a distribution of model predictions (results). The prediction distribution spread and shape then more accurately represents the reliability and precision of the results. This technique is well suited to the analysis of a computer simulation as complex as the SWMM/Water Quality package used herein.

The Models

To focus on the entire system the plume model for the Detroit River provide a single index for species concentration in the high concentration region of the river. RUNOFF, TRANSPORT, RECEIVE II, QUAL II, STPSIM II, and the plume model were required for this study.

The Variables

There are over 100 input variables for RUNOFF, TRANSPORT, RECEIVE II, QUAL II and the plume model.

In the present Monte Carlo analysis, only the dominant potential error sources were evaluated. These included the following:

RUNOFF: PCTZER, the percent impervious area with zero detention; WW(3), the subcatchment percent impervious; CBFACT(4), the catch-basin BOD concentration; RCOFF, the Pollutant Washoff Coefficient.

TRANSPORT: No changes (the pipe network is considered as fixed for a given alternative).

RECEIVE II: Headwater concentrations, Benthic deposit rate.

PLUMES: Turbulent Diffusion Coefficient.

THE METHOD

Consider the Detroit River turbulent diffusion, K_y, coefficient as an example of an input variable for the Monte Carlo Analysis.

The probability density function for K_y, assuming that it is a normally distributed function is given by:

$$f(K_y) = \frac{1}{\sqrt{2\pi}\,\sigma_{K_y}} \exp\left[-\frac{\frac{1}{2}(K_y - (\bar{K}_y)^2}{\sigma_{K_y}^{\ 2}}\right],$$

where $\bar{K}_y$ is the average turbulent diffusivity, and σ_{K_y} is the standard deviation of the distribution of K_y values.

Let $x = \frac{K_y - \bar{K}_y}{\sigma_{K_y}}$, so that $f(x) = \frac{1}{\sqrt{2\pi}} e^{-\frac{1}{2}x^2}$

The probability of finding a value of x at $x' \pm \Delta x$ is equal to $f(x')\Delta x$, since

$\int_{-\infty}^{\infty} f(x)dx = 1.$

Because of the cost of running the overall model, a limited number of runs (8) could be made. Hence, the distributions were truncated at the first standard deviation to eliminate spurious results.

If we identify $F(x) = \int_{-\infty}^{x} \frac{1}{\sqrt{2\pi}} e^{-\frac{1}{2}x^2} dx$, as the

cummulative distribution function of the standardized normal random variable, then $\int_{-1}^{1} f(x)dx = F(1) - F(-1)$ becomes the normalization constant for the more limited domain between $\pm 1\sigma$.

The probability of finding a value of x at $x' \pm \Delta x$ then becomes

$$\frac{f(x')\,\Delta x}{F(1)-F(-1)}.$$

From exponential tables, F(1) - F(-1) = 0.8413 - 0.1587 = 0.6826.

Using a system subroutine, a random number between 0 and 32767 is generated.

Then, that random number is used to select the value of the cummulative frequency function, F(x) which is between F(-1) and F(1). This weights the selection in accord with the probability density.

Now the error function is defined by

$$\text{erf}(y') = \frac{2}{\sqrt{\pi}} \int_0^{y'} e^{-y^2} dy$$

Since

$$\int_0^x \frac{1}{\sqrt{2\pi}} e^{-\frac{1}{2}t^2} dt = \frac{1}{\sqrt{\pi}} \int_0^{y'} e^{-y^2} dy$$

where $\frac{1}{\sqrt{2}} t = y$; $\frac{1}{\sqrt{2}} x = y$

then, $\int_0^x f(t)dt = \frac{1}{2}\,\text{erf}(x/\sqrt{2})$

The error function is therefore easily useable as a related measure of the cummulative frequency function, and it varies from -0.3413 to +0.3412 according to the random number selected.

Finally, the inverse error function is computed to select x, and then the random value of K_y is computed from

$$K_y = \sqrt{2}\,x\,\sigma_{K_y} + \bar{K}_y.$$

Application of the Method

The Monte Carlo method was applied to alternatives 00, 07, 11, 12, and 25 by selecting a key event (22) for each alternative and preparing cummulative frequency distributions for each alternative. These alternatives were selected because they represent a subset of best alternatives and a subset of those alternatives which have been rejected. Event 22 was chosen because, of the 32 events comprising the entire year, event 22 exhibited a total rainfall volume, an average rainfall, and an event duration characteristic of the representative values of the 1965 rain year. The event is described in Table 3.

Eight runs were made for each alternative by the random number selection process. The same random numbers were used for each alternative to provide a consistent set of variables from alternative to alternative.

TABLE 3

RAIN EVENT NUMBER	EVENT DURATION (HOURS)	TOTAL RAINFALL VOLUME (CF x 10^6)	AVERAGE RAINFALL (INCHES/EVENT)
1	35	74.25	0.216
2	33	50.35	0.146
3	49	452.0	1.313
4	48	76.86	0.223
5	44	123.8	0.360
6	36	42.98	0.125
7	70	324.5	0.942
8	29	101.0	0.293
9	107	342.6	0.995
10	53	565.1	1.641
11	29	67.52	0.196
12	27	56.2	0.163
13	26	76.54	0.222
14	41	338.8	0.984
15	45	277.1	0.805
16	28	124.6	0.362
17	30	178.3	0.518
18	60	196.2	0.570
19	91	294.0	0.854
20	84	453.6	1.317
21	68	106.1	0.308
22	55	245.0	0.712
23	30	110.1	0.320
24	52	215.5	0.626
25	38	229.5	0.667
26	84	166.3	0.483
27	85	925.7	2.688
28	31	46.81	0.136
29	42	121.7	0.353
30	42	165.8	0.482
31	79	478.6	1.390
32	63	986.9	2.866
SUM TOTAL	1634	8014.31	
AVERAGE (TOTAL 32)	51.1	250.4	0.727

TABLE 4
AVERAGE FECAL COLIFORM CONCENTRATIONS IN THE DETROIT RIVER

ALTER-NATIVE	DESIGN VALUE	MINIMUM	MAXIMUM	RANGE AS A FRACTION OF THE DESIGN VALUE
0	1.63×10^2	1.41×10^2	1.94×10^2	0.327
13	1.23×10^2	1.11×10^2	1.62×10^2	0.415
12	5.82×10^1	5.55×10^1	7.53×10^1	0.340
7	3.21×10^1	2.66×10^1	3.27×10^1	0.190
25	3.07×10^1	2.47×10^1	3.19×10^1	0.234

RESULTS AND DISCUSSION

The Detroit River

The results of the Monte Carlo sensitivity runs have been summarized using the overall cummula tive frequency distributions which are given in tabular form for the Rouge and Detroit Rivers. All eight parameters are included therein but we will focus on fecal coliform in the Detroit River and both fecal coliform and dissolved oxygen in the Rouge River.

In Table 4, we summarize the range of average fecal coliform concentrations predicted for each alternative in the Detroit River. The geometric average for the alternative was employed in the ranking procedure. The "design value" listed in these tables is the geometric average of the event 22 run, including all elements of the river, and is not necessarily equal to the full year design value. In the random selection process, this design value did not necessarily fall at the mean of the limited set.

The range of average values displayed in Table 4 represents the aggregate effect of one standard deviation input parameter uncertainties. It is not a true standard deviation of output, but within realistic time and fiscal constraints, it is a reasonable representation thereof. The average values were used in the alternative analysis. Table 5 shows that there is considerable improvement from alternative to alternative as compared to "future no action" (0) and the uncertainty is small compared to the difference predicted.

We can gain a more complete picture of the uncertainty by plotting the cummulative frequency distributions. Since in all cases over 50 percent of the data in the Detroit River represents dry weather or background concentrations, we begin that plot at the lowest percentile of interest for alternative 0, the FNA alternative.

In Figure 10, the cummulative frequency distributions for alternatives 0, 12, and 7 are displayed. The concentrations of fecal coliform are given on a logarithmic scale so that the bandwidths expand as the concentrations decrease toward background. This method displays an apparent increasing relative uncertainty at lower concentrations, despite the fact that the absolute error is actually decreasing. The solid lines connect the outer bounds on each alternative. The dotted line extrapolates toward background. There are insufficient points in this area to adequately define the transition toward background which has a very narrow uncertainty band. Using Figure 10, we can definitely say that there is a major difference between alternatives 0, 12, and 7 over the entire range of wet weather data.

In Figure 11, alternatives 0, 25, and 7 are compared. But 7 and 25 are far better than no action, but there is no clear cut difference between 7 and 25.

In Figure 12, we have plotted the cummulative frequency distributions for alternatives 0, 13, and 25. There is clearly a major difference between alternative 0 and alternative 25, but there is little difference between alternatives 0 and 13; there is some indication that alternative 13 is somewhat better than no action.

Close examination of alternatives 7 and 25 show that alternative 25 exhibits a distribution with high extreme values but with predominantly (but not exclusively) lower intermediate values. Overall, the geometric mean for alternative 25 is lower than the geometric mean for alternative 7, but for some combinations of input data, the reverse is true. Hence, the range displayed appears to be a reasonable error band. Obviously the cummulative frequency distributions cross and hence it is difficult to definitively rank these two alternatives. Similar behavior is observed for the ranking of alternatives 0 and 13.

The Rouge River

The same process was followed for the Rouge River. There is a major difference in the results, however. The Rouge River has major environmental impacts during dry weather because the headwaters are often contaminated and because of benthic demand. The latter is at least in part caused by CSO deposits which affect the dissolved oxygen concentrations long after the storm event. Thus, unlike the Detroit River which rapidly cleanses itself after the storm and which is only affected at some times and some places during the storm event, a major portion of the relevant data in the Rouge occurs during dry weather.

FIGURE 10

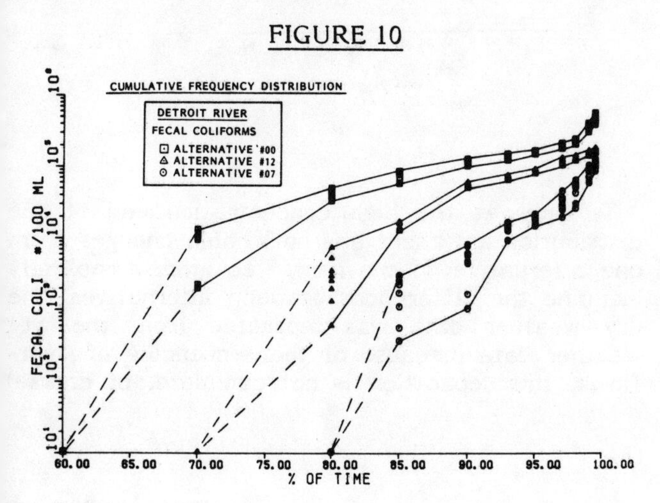

FIGURE 11

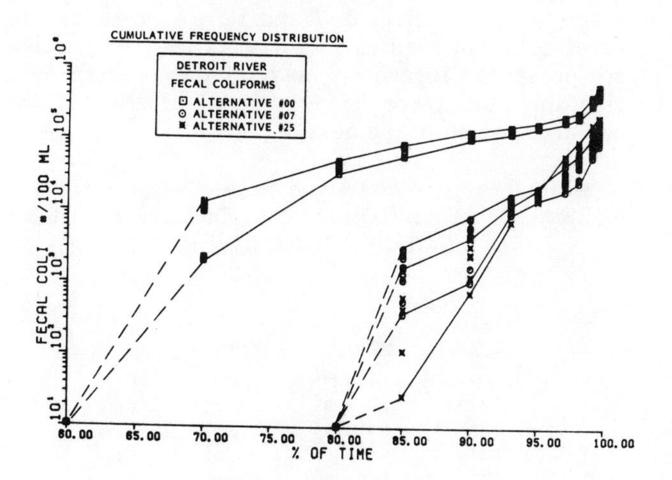

FIGURE 12

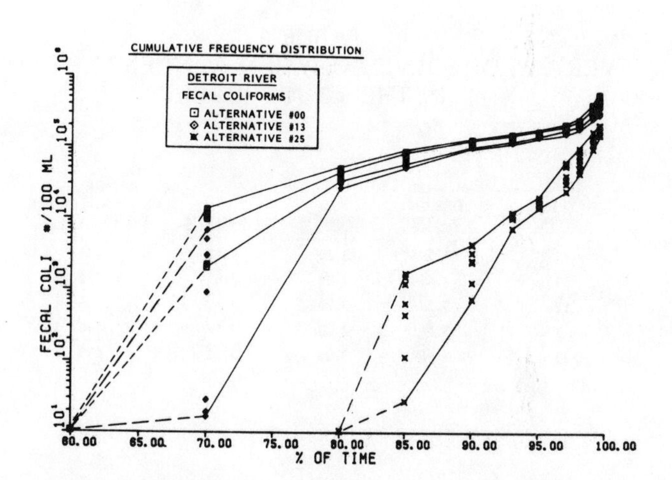

In Table 5 the Monte Carlo results for the geometric average fecal coliform concentrations are presented. For the dissolved oxygen concentration in the Rouge River the arithmetic averaged results are summarized in Table 6.

In the Rouge River for fecal coliform there is little to choose from in the design values of the alternatives. Moreover, the range of uncertainty is considerably greater than the design value or the difference between design values. The same is true for dissolved oxygen. There is little difference between alternatives and the spread in the error bound is much greater than the differences between alternatives. It is not clear that any expenditure of funds is warranted under these circumstances.

We have plotted the entire cummulative frequency distribution for the Rouge River; sample plots for alternatives 0, 7 and 12 are presented for fecal coliform Figures 13 to 15. These alternatives are presented separately because they overlap considerably and there is very little difference from one alternative to the next.

TABLE 5
AVERAGE FECAL COLIFORM CONCENTRATIONS IN THE ROUGE RIVER

ALTERNATIVE	DESIGN VALUE	MINIMUM	MAXIMUM	RANGE AS A FRACTION OF THE DESIGN VALUE
0	2.15×10^3	1.60×10^3	4.77×10^3	1.474
13	1.96×10^3	1.35×10^3	6.09×10^3	2.418
12	2.15×10^3	1.60×10^3	4.77×10^3	1.474
7	2.15×10^3	1.60×10^3	4.77×10^3	1.474
25	2.07×10^3	1.36×10^3	6.33×10^3	2.401

TABLE 6
AVERAGE DISSOLVED OXYGEN CONCENTRATIONS IN THE ROUGE RIVER

ALTERNATIVE	DESIGN VALUE	MINIMUM	MAXIMUM	RANGE AS A FRACTION OF THE DESIGN VALUE
0	3.56	3.20	3.72	.146
13	3.66	3.09	3.83	.202
12	3.56	3.20	3.72	.146
7	3.56	3.20	3.72	.146
25	3.66	3.09	3.82	.199

FIGURE 13
CUMMULATIVE FREQUENCY DISTRIBUTION FOR THE ROUGE RIVER FECAL COLIFORM CONCENTRATIONS FOR ALTERNATIVE 0

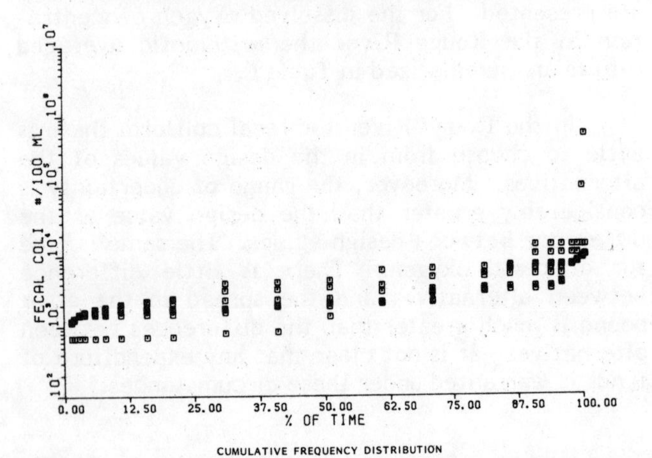

FIGURE 14
CUMMULATIVE FREQUENCY DISTRIBUTION FOR THE ROUGE RIVER FECAL COLIFORM CONCENTRATIONS FOR ALTERNATIVE 7

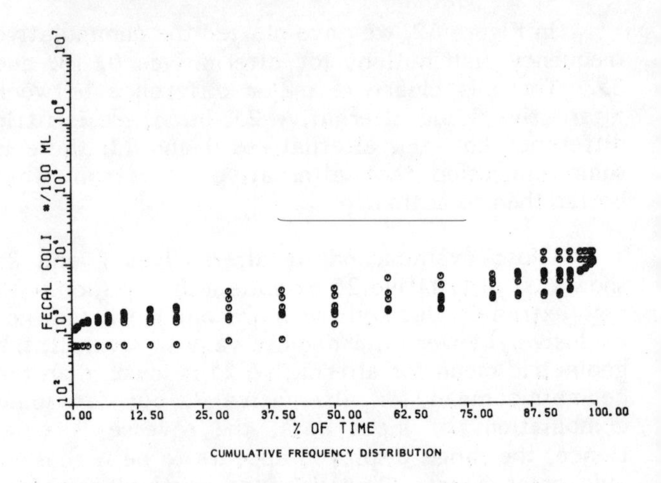

FIGURE 15
CUMMULATIVE FREQUENCY DISTRIBUTION FOR THE ROUGE RIVER FECAL COLIFORM CONCENTRATIONS FOR ALTERNATIVE 12

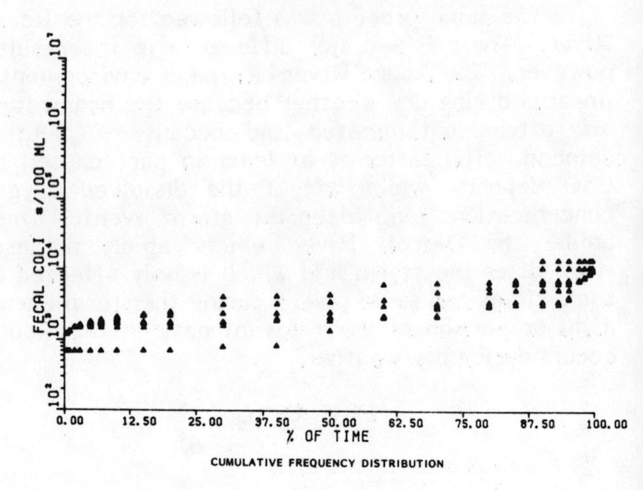

Only at the high concentration end of the distribution are there any noticeable changes from one alternative to the next. To more completely examine the differences between alternatives, the dry weather data was separated from the wet weather data (because of the sequencing of overflows, this separation is not complete, of course)

and the data examined for alternatives 0 versus 25 for two randomly selected data sets. In that calculation we do see an improvement from no action (0) to a highly treated alternative (25) for both data sets. Nonetheless, the distinction is not great and there is considerable overlap even between these alternatives for the full range of randomly chosen input data sets. Thus it is difficult to quantify the benefit of specific construction to the Rouge River using the Planning Level Models. By comparison, for the Detroit River, the environmental impacts are almost solely due to CSO's (for fecal coliform) and the benefits are clearly defined and the alternatives can be readily separated.

CONCLUSION

It is clear from the display of cummulative frequency distributions that the Monte Carlo variation of input parameter values yields a useful measure of the output uncertainty for a complex model. The uncertainty analysis performed herein allowed us to demonstrate that no action on the Rouge River, within the confines of the DWSD study area, was the only valid engineering course of action, regardless of the cost. Likewise, the uncertainty analysis demonstrated that significant benefit would be achieved by several alternatives considered for the Detroit River, allowing this portion of the program to go on to cost-benefit analysis.

LITERATURE CITED

1. International Joint Commission Great Lakes Water Quality Board, Surveillance Subcommittee Reports Appendix B for 1975, 1976, 1977, 1978, 1979 and 1980, Joint Commission, Windsor, Ontario, Canada.

2. Federal Water Pollution Control Act, Public Law 92-500, Act 33USC 1251-1376 (1972).

3. Alternate Facilities Interim (AFIR) Report, Giffels/Black and Veatch, Report on the Detroit Section 201 Study, June (1981).

4. Michigan Water Resources Commission, Bureau of Water Management, Department of Natural Resources, "Rouge River Basin: General Water Quality Survey and Stormwater Survey, June-September, 1973, March, 1974.

5. United States District Court, Eastern District of Michigan Southern Division, Civil Case Num ber 7-71100, United States of America versus City of Detroit, Detroit Water and Sewerage Department and the State of Michigan, September 14, 1977 Order, Honorable John Feikens, United States District Judge.

6. Graham, Malaise J., Water/Engineering and Management, May (1982).

7. Huber, Wayne, James P. Heaney, and W. Alan Peltz, "November 1977 Release of SWMM", presented at the SWMM Users Group Meeting, Milwaukee, (1977).

8. Anderson, H. M., "A Dynamic Simulation Model for Wastewater Renovation Systems", PhD Dissertation, Wayne State University, Detroit (1981).

9. Roesner, L., P. R. Giguere, and D. E. Evenson, "Users Manual for the Stream Quality Model", Southeast Michigan Council of Governments Report, July (1977).

10. Raytheon Company, Oceanographic and Environmental Services, "New England River Basins Modeling Project, Part I: Reciev II Water Quantity and Quality Model", EPA Contract No. 68-01-1890, December, 1974.

11. Liang, Chein-Sung, S. Winkler, and R. H. Kummler, "A Gaussian Plume Model of a Two Dimensional River", presented at the 91st National AIChE Meeting, Detroit (1981).

12. Roginski, Gregory T., "A Finite Difference Model of Pollutant Concentrations in the Detroit River from Combines Sewer Overflows", Ph D Dissertation Wayne State University, Detroit (1981).

13. Anderson, H. M., "Combined Sewer Overflow Modeling: STPSIM2, A Dynamic Model of the Wastewater Treatment Plant", presented at the 91st National AIChE Meeting, Detroit (1981).

14. Anderson, J. A., C. D. Harlow, and J. Baranec, "Combined Sewer Overflow Modeling in the Detroit 201 Study Using SWMM", presented at the 91st National AIChE Meeting, Detroit (1981).

15. Anderson, J. A., C. D. Harlow, J. Baranec, and H. M. Anderson, "Combined Sewer Overflow Modeling in the Detroit 201 Study using SWMM", Proceedings of the USEPA Stormwater Management Model User's Meeting, Austin, January (1981).

16. El Sharkawy, Alaa, "Water Quality Modeling for One-Dimensional Rivers", Master's Thesis Wayne State University, Detroit (1981).

17. El Sharkawy, Alaa and Esmail Jamshidi, "A Comparison of an Analytical, One Dimensional River Simulation with Numerical River

Models", presented at the 91st National AIChE Meeting, Detroit (1981).

18. Filardi, Raul and Barbara Harvey-Brayton, "CSO Control Alternatives Analysis", presented at the 91st National AIChE Meeting, Detroit (1981).

19. Harlow, C. D., E. Jamshidi, R.H. Kummler, J. G. Frith and J. A. Anderson, "One Dimensional Water Quality Models for Dynamic, Small Rivers: The Rouge River", Proceedings of the USEPA (SWMM) Stormwater Management Model User's Meeting, January (1981) Austin.

20. Harlow, C. D., and J. Frith, "Numerical Water Quality Models for One Dimensional Rivers", presented at the 91st National AIChE Meeting, Detroit, August (1981).

21. Harvey-Brayton, Barbara L., and Raul E. Filardi, "Combined Sewer Overlow Abatement Alternative Development", presented at the 91st National AIChE Meeting, Detroit, August (1981).

22. Kummler, R. H., G. Roginski, C-S. Liang, S. Winkler and J.A. Anderson, "Two Dimensional Water Quality Models for Dymanic, Large Rivers: The Detroit River", Proceedings of the USEPA (SWMM) Stormwater Management Model User's Meeting, January (1981), Austin.

23. Kummler, R. H., "SWMM Modeling for the Detroit 201 Final Facilities Plan: Final Results", Proceedings of the USEPA SWMM User's Meeting, October, Ottawa, Ontario, (1982).

24. Liang, Chein-Sung, "Use of Multispectral Remote Sensing Data to Predict the Turbulent Diffusion Coefficient in the Detroit River", Master's Thesis Wayne State University, Detroit (1981).

25. Roginski, G. and R. H. Kummler, "A Finite Difference Model of a Two Dimensional River", presented at the 91st National AIChE Meeting, Detroit (1981).

26. Upmeyer, David W., "The Alternatives Analysis Procedure", presented at the 91st National AIChE Meeting, Detroit (1981).

27. Giffels/Black & Veatch, Segmented Facilities Plan prepared under Section 201 Facilities Planning Grant for the Detroit Water and Sewerage Department, January (1978).

28. Southeast Michigan Council of Governments, "1975 Land Use Inventory", June (1978).

INTERVAL ESTIMATES OF PROJECTED COST

This paper presents two topic areas. First, a mathematical model to approximate the probability distribution of future cost is presented. Although the model has been used in some applications, it is new in the area of cost forecasting. Second, the means of supplying the parameters in this model are investigated. Subjective judgement has been resorted to previously and is the method used herein. However, the points used are the quartiles (25%, 50%, 75%) instead of most likely, pessimistic, and optimistic. After justifying these procedures, their deficiencies are noted and means for minimizing them are discussed. A comprehensive example is presented.

PAUL R. DUNLAP

U.S. Environmental Protection Agency
Industrial Environmental Research Laboratory
Research Triangle Park, NC

SCOPE

In the recent past, large errors have occurred in forecasting the cost of projects. Investigations (1) as to the cause of these errors revealed two broad types. First, the majority of cost overruns were the results of improper cost functions (viz., one or more variables affecting the cost were not included in the cost function). Second, and not as large as improper cost functions, was inaccurate forecasting of some of the variables in the cost function. This paper deals with the second type of error.

There has been considerable work done on this problem of inaccurate forecasting, but the results have not been widely applied (2, 3, and 4). There are two reasons for this. First, although several authors use similar approaches they are inconsistent in their application. Second, there is the off-hand attitude that a computer simulation will somehow solve the problem of inaccurate forecasts. However, a computer program is simply a rigorous statement of an approach and, as such, does not remove inconsistencies, but only disguises them. We have looked into the inconsistencies in various authors' approaches and attempted to minimize or eliminate them. The exploration, examination, and resolution of this investigation is the principal subject of this paper.

CONCLUSIONS AND SIGNIFICANCE

An approximation of the probability distribution of cost can be obtained by use of a Taylor series and the central limit theorem (5, 6, and 7). An investigation of this approximation has shown it to be very good. In fact, this approximation requires less input information than a computer simulation, and permits the same probability statements about costs. This approximation requires that only the mean and variance of each variable in the cost function be known. By comparison, a Monte Carlo procedure (computer simulation) requires that the probability distribution of each variable be developed.

The problem of supplying these parameters when they are unknown was investigated. Since there is no universe from which to sample, there is no solution which will satisfy a person who is a strict related frequentist. However, if one accepts subjective probability, then we can proceed. Until now normal practice has been to obtain subjective judgment of three points on each cost curve: the most likely, an optimistic one, and a pessimistic one. From these three points the mean and variance of each cost curve were determined. However, the inconsistencies mentioned above arose as a result of different definitions of optimistic and pessimistic which, in truth, caused

---

On leave from Ohio University, Athens, OH.

0065-8812-82-6282-0220-$2.00

different formulas to be used to obtain the mean and variance. Although there appears to be a best definition and formula (3, 8) to compute the mean and variance, these three points are not the logical ones to use for subjective probabilities since they are not compatible with the apparent mental processes exercised in subjective judgment. As a result of our investigation, we have recommended that the three points be the quartiles (i.e., 25%, 50%, and 75% points). These three points are chosen since they are more amenable to "a step-wise program" which guides and enhances the judgment process.

From the probability distribution that is developed, an interval statement can be made (e.g., the probability is α that the cost will be less than k dollars). This interval gives more information than the present one-point estimate, thus reducing the uncertainty as well as quantifying it.

THE MODEL

The future cost y of a system is a function of many cost variables:

$$y = g(x_1, x_2, \ldots, x_i, \ldots, x_n) \quad (1)$$

which we shorten to $g(x_i)$. It is reasonable to assume that many of these are random variables; therefore, y will be a random variable with distribution F(y) or a probability density function f(y). If we can determine f(y), the probability interval of cost is simply:

$$P(a < y < b) = 1 - \alpha = \int_a^b f(y)\,dy \quad . (2)$$

Therefore we seek f(y).

If the function g is a sum and if n is large enough, the central limit theorem tells us that F(y) is approximately normally distributed. If g is a product then F(ln y) is approximately normal. However, if g is some combination of products and sums (which frequently is the case), what then is the density of f(y)? The answer for practical purposes (without proof) is also a normal density. A complex problem frequently arises when attempting to find the mean and variance for this type function.

To simplify this problem much of the previous work made simplifying assumptions which introduced large errors. First, if the x_i's are assumed to be independent, this problem becomes more tractable. However, in the real world this assumption is suspect. Second, when products were involved, logs were taken and the mean and variances found. The antilogs of these were then, more or less indiscriminately, combined with the means and variances of terms not involving products. Combining medians and means is the mathematical equivalent to adding apples and oranges. In short, the simplifying assumptions previously used are too rough and indiscriminant.

If the individual x_i's are random variables with finite variances, a Taylor series expansion, evaluated at $\mu_1, \mu_2, \ldots \mu_n$, will approximate $y = g(x_i)$, or:

$$y = g(\mu_i) + \sum_{i=1}^{n} g_i'\,(\mu_j)\,(x_i - \mu_i)$$

$$+ \frac{1}{2}\sum_{i=1}^{n} g_i''\,(\mu_j)\,(x_i - \mu_i)^2 \quad (3)$$

$$+ \sum_{1}^{n} g_{ij}''\,(\mu_k)(x_i - \mu_i)(x_j - \mu_j)$$

where

g_i' is the partial derivative of $g(x_i)$ with respect to x_i.

g_i'' is the second partial derivative of $g(x_i)$ with respect to x_i.

g_{ij}'' is the second partial derivative of $g(x_i)$ with respect to x_i and then with respect to x_j.

Equation (3) is a sum of more than n terms; therefore, f(y) should be approximately normal with its mean equal to the expected value of Equation (3) and its variance equal to the variance of Equation (3).

The expected value of Equation (3) is:

$$E(y) = \mu_y = g(\mu_i) + \frac{1}{2}\sum_{i=1}^{n} g_i''\,(\mu_j)\,\sigma_i^2 + \Sigma g_{ij}'' \cdot (\mu_k)\,\mathrm{cov}\,x_i x_j + R \quad (4)$$

This equation clearly shows the error introduced by an incorrect assumption of independence of the x_i's.

The variance of Equation (3) can be found by determining $E(y^2)$ from Equation (3) and then subtracting $[E(y)]^2$ from it. The algebra becomes very messy when $n > 3$. However, Hald (6) gives an approximation for this variance:

$$\text{Var}(y) = \sigma_y^2 = \sum_{i=1}^{n} [g_i(\mu_j)]^2 \sigma_i^2$$

$$+ 2\sum_{i=1}^{n} g_i'(\mu_k) \tag{5}$$

$$\cdot\ g_j'(\mu_\ell) \text{ cov } x_i x_j$$

$$k \neq i \qquad 1 \neq j$$

Example

$y = x_1 x_2 + x_3/x_4$

$g'x_1 = x_2 \qquad g''x_1 = 0 \qquad g''_{12} = 1$

$g'x_2 = x_1 \qquad g''x_2 = 0 \qquad g''_{13} = g''_{14} = 0$

$g'x_3 = 1/x_4 \qquad g''x_3 = 0 \qquad g''_{23} = g''_{24} = 0$

$g'x_4 = -x_3/x_4^2 \quad g''x_4 = 2x_3/x_4^3 \quad g''_{34} = -1/x_4^2$

$$E(y) = \mu_1\mu_2 + \mu_3/\mu_4 + (\mu_3/\mu_4^3)\sigma_4^2$$

$$+ \text{ cov } x_1x_2 - 1/\mu_4^2 \text{ cov } x_3x_4$$

and

$$\text{Var }(Y) = \mu_2^2\,\sigma_1^2 + \mu_1^2\,\sigma_2^2$$

$$+ (1/\mu_4)^2\,\sigma_3^2 + (\mu_3^2/\mu_4^4)\,\sigma_4^2$$

$$+ 2\,\mu_1\,\mu_2 \text{ cov } x_1x_2$$

$$+ 2\,\mu_2/\mu_4 \text{ cov } x_1x_3$$

$$- 2\,\mu_2\,\mu_3/\mu_4^2 \text{ cov } x_1x_4$$

$$+ 2\,\mu_1/\mu_4 \text{ cov } x_2x_3$$

$$- 2\,\mu_1\mu_3/\mu_4^2 \text{ cov } x_2x_4$$

$$- 2\,\mu_3/\mu_4^3 \text{ cov } x_3x_4$$

This result requires three different approximations (Equations (3) and (5) and the central limit theorem). The question of magnitude of error introduced has been investigated. On a large variety of cases the results of this procedure agreed to three significant figures with a computer simulation of 10,000 trials.

Example

Consider the cost of a system which is the sum of three subsystems. The cost for each subsystem includes materials and labor. The overall cost also includes overhead, engineering fee, contractor's fee, freight, spares, shakedown, contingencies, and interest during construction.

The cost equation is

$$y \doteq \{\,[(1+x_1)(1+x_2+x_3+x_2x_3) + x_4+x_5+x_6]\,[1+x_7]\,[1+x_8x_9]\,\}$$

$$\cdot\ \{5.7\,x_{10} + 9.4\,x_{11} + x_{12}\}$$

where the variables and their means, variances, and correlations are given in Table 1.

TABLE 1. PARAMETERS OF COST VARIABLES

x_i	μ	σ^2
x_1 Overhead	0.24	0.01416
x_2 Contractors' fees	0.10	0.005495
x_3 Engineering fees	0.099	0.00004275
x_4 Freight	0.05	0.00004651
x_5 Shakedown	0.05	0.00002693
x_6 Spares	0.01	0.00001163
x_7 Contingency	0.20	0.001163
x_8 Interest rate	0.16	0.00009769
x_9 Construction time (months)	0.375	0.03435
x_{10} Subsystem 1	146.0	226.0
x_{11} Subsystem 2	111.0	130.3
x_{12} Subsystem 3	167.0	295.8

Correlation Matrix, ρ_{ij}

$\rho_{1\ 10} = \rho_{1\ 11} = \rho_{1\ 12} = 1.0$

$\rho_{4\ 10} = \rho_{4\ 11} = \rho_{4\ 12} = 0.8$

$\rho_{6\ 10} = \rho_{6\ 11} = \rho_{6\ 12} = 0.9$

$\rho_{8\ 9} = -0.6$

Covariance Matrix

x_i/x_j	x_9	x_{10}	x_{11}	x_{12}
x_1		1.789	1.358	2.047
x_4		0.08200	0.06221	0.09370
x_6		0.04614	0.03503	0.05279
x_8	-0.001099			

The approximation of y, μ_y, and σ_y^2 requires derivatives with respect to the x_i's. These are given in Table 2.

Using Equations (4) and (5), we obtain E(y) = 4220 and Var(y) = 498680.

Because of the central limit theorem, y is approximately normal. Plotting a normal curve with mean = 4220 and variance = 498680 gives Figure 1. From this we may obtain any probability we wish, such as:

$$P\ (y < 5380) = 0.95$$

or

$$P\ (3060 < y < 5380) = 0.90$$

Obviously this probability distribution of future cost gives more information than the one-point value that is generally given. In addition, analysis of Equations (3), (4),

TABLE 2. PARTIAL DERIVATIVES OF COST FUNCTION

i	g_i'
1	$3141(x_i - 24)$
2	$3541(x_2 - 0.1)$
3	$3544(x_3 - 0.009)$
4	$2598(x_4 - 0.05)$
5	$2598(x_5 - 0.05)$
6	$2598(x_6 - 0.01)$
7	$3484(x_7 - 0.20)$
8	$1479(x_8 - 0.16)$
9	$631.0(x_9 - 4.5/12)$
10	$11.67(x_{10} - 146)$
11	$1924(x_{11} - 111)$
12	$2.470(x_{12} - 167)$
$g_i'' = 0$ for all i	
$g_{1\ 10}' = 8.765$	
$g_{1\ 11}' = 14.47$	
$g_{1\ 12}'' = 1.538$	
$g_{4\ 10}'' = g_{6\ 10}'' + 7.250$	
$g_{4\ 11}'' = g_{6\ 11}'' = 11.96$	
$g_{4\ 12}'' = g_{6\ 12}'' = 1.272$	
$g_{8\ 9}'' = 3944$	

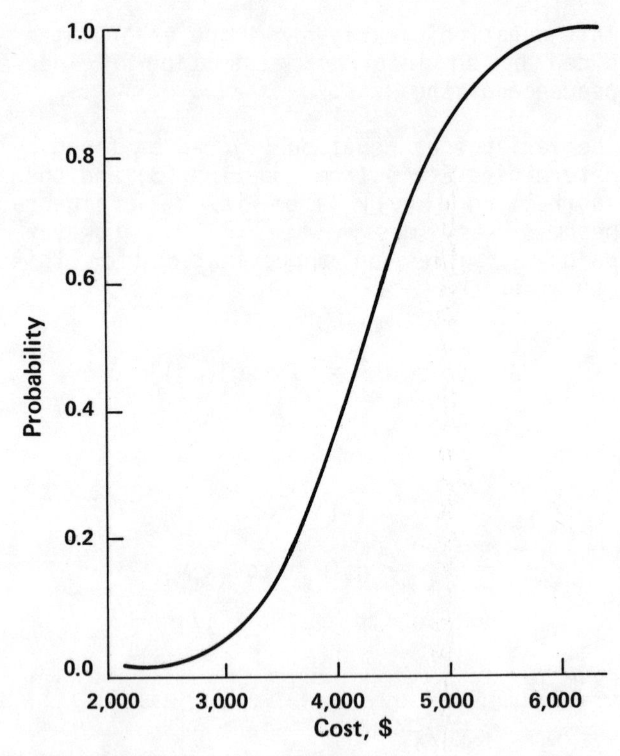

Figure 1. Distribution of project cost.

and (5) shows that it is better information because it is unbiased. The general procedure for giving a point value for future cost is to compute the first term of Equation (3). From Equation (4), it is seen that this value is too small if any variable has a finite second derivative or is positively correlated. In addition, the expected value of the second term in Equation (3) will not equal zero if one takes logarithms.

It should also be emphasized that this distribution is the same (except for trivial approximation errors) as produced by a computer simulation (Monte Carlo). In fact, the derivatives of Table 2 provide a preliminary sensitivity analysis and also serve as a good review of cost function $g(x_i)$. Contrary to computer system advocates' claims (9), a computer simulation is not a panacea; a critical review of the cost function is vital. This cost function review is an inherent benefit of this model.

However, the use of either this model or Monte Carlo requires that the means, variance, and correlation (given in Table 1) be known. This raises the question: "What does one do when the distribution or the parameters of the x_i's are unknown?"

UNCERTAINTY ANALYSIS

The statistical answer of estimation from sampling is not an appropriate answer to the question posed here because there is no universe from which sampling may be done. Therefore, some other approach is necessary because values of these parameters are essential if a cost forecast is to be done. There are several techniques (locked door, consensus, etc.) which are used out of necessity. All of these use subjective judgments to provide the parameters. A literature search revealed that a PERT approach with minor modifications was the method used to deal with these judgments in a methodical manner.

This approach used a three-point guess of the cost for each x_i. A cost engineer was asked to provide for each x_i:

1. The most probable value, C_m.
2. A pessimistic value, C_p.
3. An optimistic value, C_0.

Copying the work of PERT (10)

$$E(x_i) = \mu_i = \frac{4\,C_m + (C_o + C_p)}{6} \quad (6)$$

and variance x_i

$$\sigma_i{}^2 = \{C_o - C_p)/6\}^2 \quad . \quad (7)$$

Although using the same approach, different authors have developed variations of these equations. In the equation for variance, Holland et al. (4) divide by 8; Ferencz (2) divides by 5.2; while Moder and Rogers (3) use 3.2. Clearly the reason for these different values is in the authors' definitions of pessimistic and optimistic. These differences can be translated into the proportion of the variable that lies between the pessimistic and optimistic value: Malcolm et al. (10) give no specific proportion, but $\pm 3\sigma$ has been used since the 1940's; Holland et al. (4) use 99.99%; Ferencz (2) uses 99%; and Moder and Rogers (3) use 90%. These differences indicate a need to be specific in the definition of pessimistic and optimistic. Because not only the authors but the cost engineers whose subjective judgments are being sought will also have different definitions, it is suggested that the words themselves not be used, but that specific points be requested. Because of the relationship of standard deviation to the interval being sought, the range from 5% to 95% point appears to be best (3, 8). However, because of the difficulty in dealing with large to very large probabilities, these points seem inappropriate. For example, when one reads the weather report, can one really distinguish between an 80%, 90%, and 99% chance of rain? To circumvent this problem we change the question to provide a 50% probability, just as bookies give a point spread in sporting events to provide an even-money bet.

Following the lead of Raiffa and Schlaifer (11), we ask the cost engineer to tell us the median cost. Another way of explaining this median to a cost engineer is to ask him to pick a point which partitions the cost into two segments which are equally likely. Then the investigator will bet even money on the segment of his choice and the cost engineer must bet on the other segment. This gives the median which is the point suggested instead of the most likely value or mode. To get the second and third points, use conditional probability. Assume that the cost engineer lost the first bet and offer him a chance to get even. He is to divide the given segment into two parts which are equally likely.

Then, as in the first bet, the investigator will choose the part to bet on and the engineer must bet on the other. In a similar manner, assume the cost engineer won the first bet and the investigator wants a chance to get even. This procedure has given the quartiles (i.e., 25%, 50%, 75%) and all the information needed for symmetric distributions.

If the absolute distance between the median and low value is equal to the absolute distance between the median and the high value, the median equals the mean, and the distribution is symmetric. To determine the variance from the distance between the quartiles, an assumption about the distribution is needed: the two extremes are normal and uniform.

If uniform:

$P_{0.75} - P_{0.25} = \frac{1}{2}$ R'and

$$\sigma^2 = \frac{R^2}{12} = \frac{(P_{0.75} - P_{0.25})^2}{3} \quad (8)$$

where

$$R = x_{max} - x_{min}$$

If normal:

$P_{0.75} - P_{0.25} = 1.349\ \sigma$ and

$$\sigma^2 = \frac{(P_{0.75} - P_{0.25})^2}{1.820} \qquad (9)$$

There are several ways to choose between these two. In general, the normal gives a broader (conservative) interval.

If the density of x_i is skewed, the problem becomes more complex. It is suggested that more points be obtained from the cost engineer and some curve fitting be done. These points should also be developed on the even-bet basis. A few examples will show how this can be done.

Example

The results of an interview with cost engineer A are given in Table 3. The values given in Table 1 were obtained from these values, using Equation (9) for the x_i's which were symmetric.

Of the 12 variables in the problem, only 2 (x_i, overhead; and x_3, engineering fees) were considered skewed by the cost engineer who was responsible for the study. It was pointed out to him that x_1, overhead, was skewed and he should reconsider it. The next day, if he still felt that it was skewed, he would be asked to give three more points: the most likely, the 12.5% point, and the 87.5% point. He was also told that the variable x_3, engineering fees, was selected as a variable by him but when he gave values he called it a constant. The values he gave the next day for x_1 were:

$x_1 < 0.1$ with probability equal .125, and

$x_2 < 0.4$ with probability equal .875.

The most likely value of $x_1 = 0.15$.

These were plotted to give Figure 2.

By working backwards we obtained the following frequency distribution from Figure 2 for variable x_1:

TABLE 3. ENGINEER A ESTIMATES

Variable	Quartiles 50%	25%	75%	Skewed
x_1	0.175	0.15	0.25	yes
x_2	0.10	0.05	0.15	
x_3	0.10	0.10	0.10	
x_4	0.05	0.0454	0.0546	
x_5	0.05	0.0465	0.0535	
x_6	0.01	0.0077	0.0123	
x_7	0.20	0.177	0.223	
x_8	0.16	0.1533	0.1667	
x_9	0.375	0.25	0.50	
x_{10}	146.0	135.9	156.1	
x_{11}	111.0	103.3	118.7	
x_{12}	167.0	155.4	178.6	

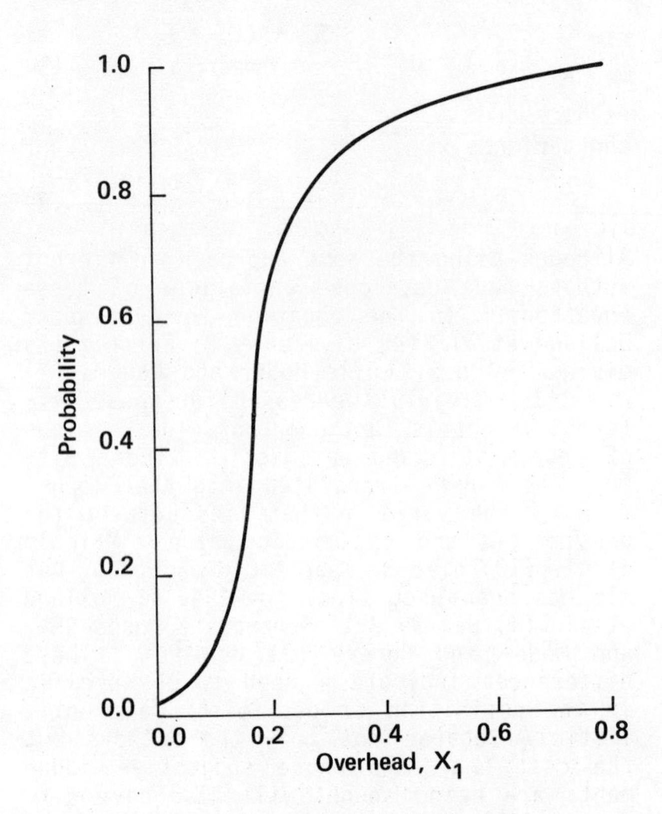

Figure 2. Distribution of overhead by Engineer A.

x_1	freq.	
5	3	
10	7	
15	15	
20	39	
25	12	$E(x_1) = 24.0$
30	7	
35	5	
40	3	$\text{Var } x_1 = 0.01416$
45	3	
50	3	
55	2	
60	1	
65	1	

Then x_3 was discussed. He said that he meant what he said. First, it was a variable, but over 50% of the time the value was 0.1. However, occasionally the price might be raised or lowered. When asked to be specific, he gave the distribution:

x_3	pb	
0.08	0.05	$E(x_3) = 0.099$
0.09	0.15	
0.10	0.70	$\text{Var } x_3 = 0.00004275$
0.11	0.10	

There remains the question of interpreting these subjective probabilities or measuring the quality of the "guesstimation". There is no known way to quantify exactly subjective probability (i.e., the variance of the subjective distributions between different individuals). However, there has been considerable research into this area.

A literature search revealed these interesting yet obvious facts:

1. Two or more heads are better than one (although not directly proportional).
2. Frequently it is impossible to come up with a consensus.

Dalkey and Helmer (12) have written extensively on the Delphi method of eliciting "expert opinions." This procedure requires several independent iterations of the expert opinions. Although at each iteration the different individuals are given feedback as to where they differ, they never meet as a group. This is to eliminate the effect of a dominant personality and still permit the exchange of information.

In collecting data for our example, we used a modified Delphi method. We used two independent cost engineers and three sessions with each. In the feedback sessions two types of information were given to them: (1) inconsistencies in the estimates they gave us, and (2) major differences with the other estimator. It was observed that both readily changed if the data were inconsistent, but were not inclined to change because of differences pointed out to them.

Table 4 gives the estimates of the second cost engineer. It lists two values for variables x_4 and x_7. These were his first and last estimates. For variable x_4, it was pointed out to the engineer that the median and mode were equal, which is inconsistent with a skewed continuous function. In addition he was told that the 5 point should give a sigmoid curve and his 5 points were off. The next day he gave the last set of data. When plotted, this produced a curve with no large inconsistencies (see Figure 3).

For variable x_7, his points were plotted (see Figure 4). There are two obvious inconsistencies. First, the 0.75 point gave a bad curve. Second, his mode was on the wrong side of the median. For a continuous unimodal function the mode, median, and mean will always be in that order going from the

TABLE 4. ENGINEER B ESTIMATES

Variable	0.50	0.25	0.75	0.12½	.87½	Mode
x_1	0.20	0.12	0.28			
x_2	0.12	0.08	0.16			
x_3	0.08	0.04	0.12			
x_4 first	0.03	0.025	0.033	0.024	0.0335	0.03
x_4 last	0.025	0.02	0.035	0.01	0.04	0.022
x_5	0.05	0.03	0.07			
x_6	0.01	0.005	0.015			
x_7 first	0.12	0.07	0.30	0.04	0.32	0.15
x_7 last	0.12	0.07	0.20	0.04	0.32	0.10
x_8	0.14	0.11	0.17			
x_9	4.0	3.0	5.0	2.0	7.5	3.5
x_{10}	145.0	130.0	160.0			
x_{11}	110.0	100.0	120.0			
x_{12}	160.0	145.0	175.0			

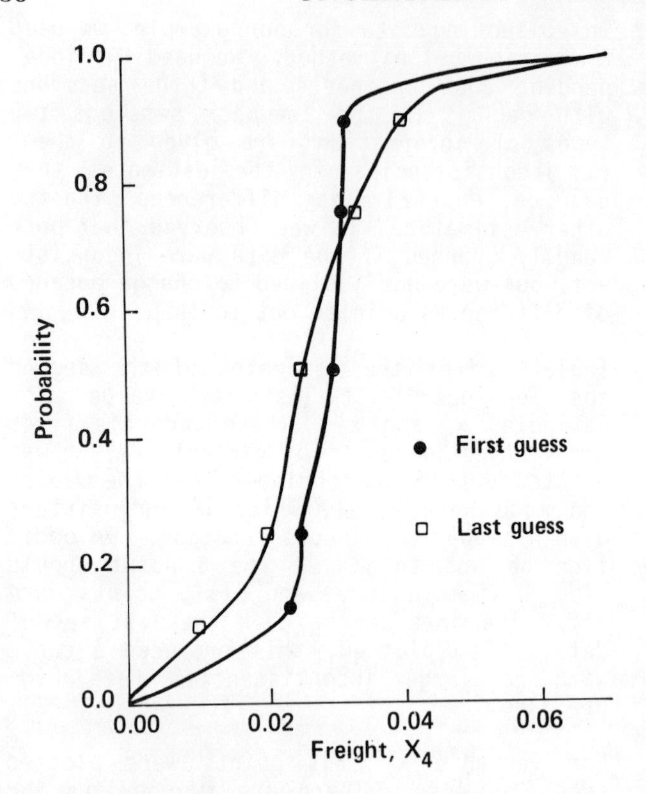

Figure 3. Distribution of freight by Engineer B.

short side to the elongated side. The next session produced the last set of values for x_7. Figure 5 is a plot for variable x_9.

From Table 4 and Figures 3, 4, and 5, we computed the mean and variances as described before. These values are presented in Table 5. From these we found f(y): E(y) = 3932 and Var (y) = 678391. Figure 6 presents F(y) for both cost engineers.

If we know (certainty) a probability curve, then the risk (chance or probability) of a cost can be determined for the distribution. When we have more than one curve and don't know which is correct (uncertainty), we need some means of combining them. Winkler (13) has investigated ways of doing this. At present a weighted average appears to be best, even if the weights are subjective. In the past it has been recommended that subjective judgments be used to create a curve. The fact that only one curve is shown is misleading because it is actually one of many. Therefore, it is recommended that more than one curve be developed.

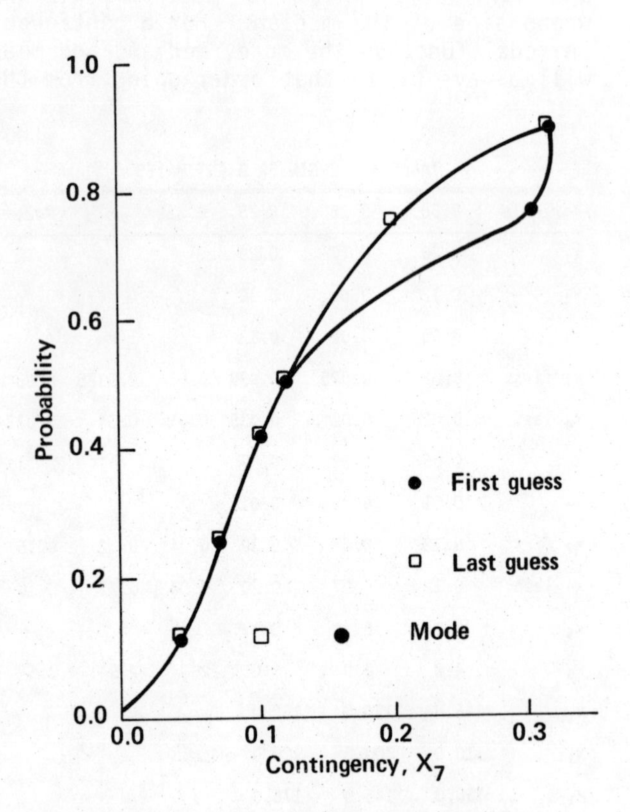

Figure 4. Distribution of contingency by Engineer B.

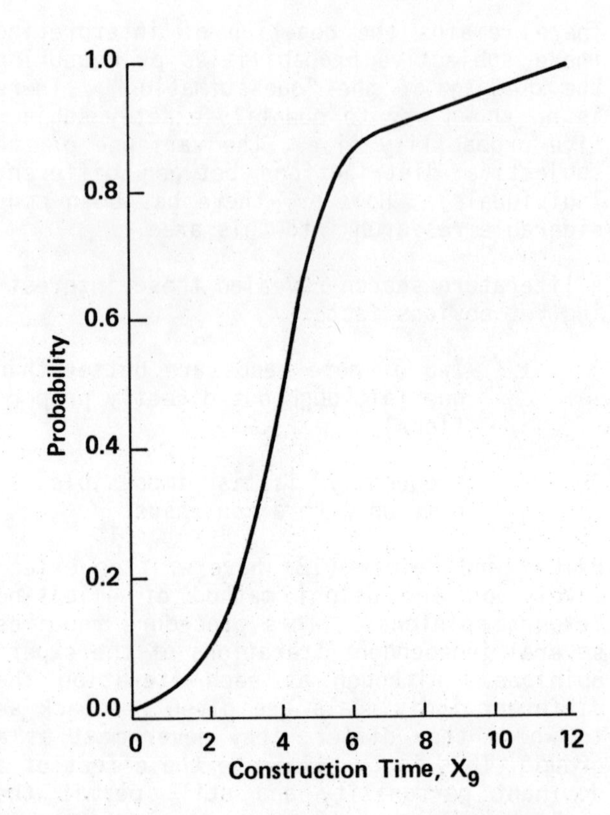

Figure 5. Distribution of construction time by Engineer B.

TABLE 5. ENGINEER B MEANS, VARIANCE, CORRELATION

Variable	μ	σ^2
x_1	0.20	0.01407
x_2	0.12	0.003517
x_3	0.08	0.003517
x_4	0.028	0.0001185
x_5	0.05	0.0008792
x_6	0.01	0.00005495
x_7	0.181	0.008791
x_8	0.14	0.001987
x_9	0.402	0.03056
x_{10}	145.0	494.6
x_{11}	110.0	219.8
x_{12}	160.0	494.6

Correlation Matrix, ρ_{ij}

$\rho_{1\ 10} = \rho_{1\ 11} = \rho_{1\ 12} = 0.9$

$\rho_{4\ 10} = \rho_{4\ 11} = \rho_{4\ 12} = 0.9$

$\rho_{6\ 10} = \rho_{6\ 11} = \rho_{6\ 12} = 0.8$

$\rho_{8\ 9} = 0$

Covariance Matrix

x_i/x_j	x_{10}	x_{11}	x_{12}
x_1	2.374	1.583	2.374
x_4	0.2179	0.1452	0.2179
x_6	0.1319	0.0676	0.1319

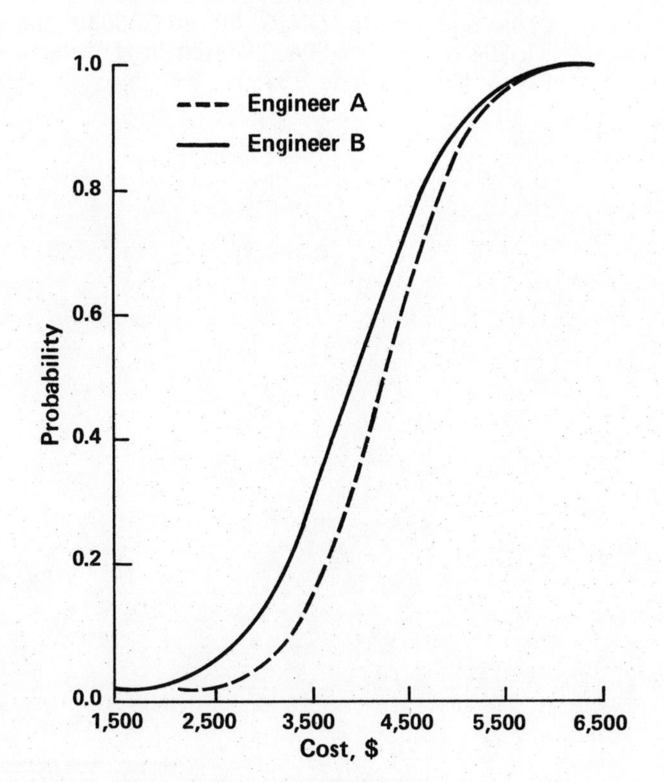

Figure 6. Comparison of cost distributions.

To illustrate, consider the above example. It was a study estimate and it took about 8 man-weeks to develop the cost function and an expected cost. According to Uhl (14) this should be within ±30%. (He gives no probability.) To develop a subjective cost probability distribution took about 3 days. Each cost probability function gives an independent review of cost function $g(x_i)$ and a cursory sensitivity analysis of $g(x_i)$ in addition to giving a probability of the cost. The fact that the ±30% of Engineer A was in agreement with this distribution added some confidence. When Engineer B's independent subjective judgments produced a curve which was in close agreement with the first curve, some more of the uncertainty was removed.

RECOMMENDATION

Since the addition of a subjective cost probability function requires little effort over and above that required to produce a point estimate, it is recommended that it be done. As the detail of the cost analysis is increased it is recommended that the number of independent cost probability distributions be increased to reduce the uncertainty involved in the "guesstimation" process.

NOTATION

$$E(x) = \int_{-\infty}^{\infty} x\, g(x)dx = \text{mean } \mu.$$

$$E(x^2) = \int_{-\infty}^{\infty} x^2 g(x)dx.$$

$$Var(x) = E(x^2) - [E(x)]^2.$$

$g'(x_i)$ = partial derivative of $g(x_1 \ldots x_i \ldots x_n)$ with respect to x_i.

$g'(\mu_j)$ = partial derivative of $g(x_i)$ evaluated at all μ_j, $i \neq j$.

$g''_{ij}(\mu_k)$ = the second partial of $g(x_i)$ evaluated at all μ_k for $j \neq i \neq k$.

covariance = $E(x_i - \mu_i)(x_j - \mu_j)$.

α = a small probability

μ_i = mean of the ith variable = $E(x_i)$.

σ^2_i = variance of the ith variable = Var (x_i).

σ_{ij} = covariance of x_i and x_j.

ρ_{ij} = correlation between x_i and x_j.

y ≡ system cost = $g(x_1, x_2, \ldots x_i, \ldots x_n)$.

F(y) ≡ y is random variable with probability distribution, F(y).

f(y) ≡ the probability density function of the derivative of F(y).

LITERATURE CITED

1. Tyler, C., Chem. Eng. 60(1), 198 (1953).
2. Ferencz, P., Chem. Eng., 59(4), 143 (1952).
3. Moder, J. J., and E. G. Rogers, Management Science, 15 (2) (1968).
4. Holland, F. A., F. A. Watson, and J. K. Wilkinson, Chem. Eng. Refresher, Parts 8-11 (1973).
5. Mood, A. M., F. A. Grayhill, and D. C. Boes, Introduction to the Theory of Statistics, McGraw-Hill Book Co., New York (1974).
6. Hald, A., Statistical Theory With Engineering Applications, John Wiley & Sons, Inc., New York (1952).
7. Kendall, M., and A. Stuart, The Advanced Theory of Statistics, Volume 1, MacMillan Publishing Co., Inc., New York (1977).
8. Pearson, E. S., and J. W. Tukey, Biometrika, 2 (1965).
9. Ross, R. C., Chem. Eng. (September 1971).
10. Malcolm, D. C., J. H. Roseboom, E. Clark, and W. Fozar, Operation Research, 7(5) (1959).
11. Raiffa, H., and R. Schlaifer, "Applied Statistical Decision Theory," Harvard University, Division of Research, Boston (1961).
12. Dalkey, N., and O. Helmer, Management Science, 9(9) (1962).
13. Winkler, R. L., Management Science, 15 (2) (October 1968).
14. Uhl, V. W., "A Standard Procedure for Cost Analysis of Pollution Control Operations, Volume I. User Guide; Volume II. Appendices," EPA-600/8-79-018a and -018b (NTIS PB 80-108038 and -108046), U.S. EPA, Research Triangle Park, N.C. (June 1979).

CORRECTING FOR MEASUREMENT ERROR: AN ANALYSIS APPLIED TO CHEMICAL PARAMETERS IN AQUEOUS MEDIA*

The accurate measurement of the concentration, C, of a chemical constitutent in effluent streams is very important in order to assure that facility processes and pollution abatement measures are working properly, that a facility is meeting regulation criteria, and that unnecessary expenditures on pollution abatement systems and operation are avoided. Unfortunately, these measurements are subject to various sources of variability, some causing significant error. As actual data tend to be skewed (not symmetric), the usual statistical model may be inappropriate. A comparison of a standard model with our proposed model shows significant differences in the reported test statistics for the concentrations. This may be critical for meeting EPA and other guidelines.

ALAN GLEIT

Versar Inc.
6621 Electronic Drive
Springfield, VA

Manufacturing facilities in general discharge waste materials resulting from their manufacturing processes. Certain chemicals in these wastes may be harmful to wildlife, plants, or man. Therefore, their disposal or release has been regulated by federal, state, and local agencies. Specific regulations regarding disposal of wastes are, however, relatively recent, and the list of regulated compounds is constantly being increased. Hence, there are numerous waste disposal sites in our nation predating these regulations which contain hazardous materials.

Active waste disposal sites are easily located, and the companies currently using them readily identified. Corrective action may then be carried out by the responsible parties. On the other hand, we face the enormous task of discovering, investigating, and remedying abandoned, inactive, and otherwise uncontrolled hazardous waste sites such as the Love Canal site in Niagara Falls, New York. Preliminary estimates indicate that there are 30,000 to 50,000 such sites throughout the nation requiring some amount of enforcement or remedial action to contain and eliminate the hazard. In connection with this investigative process, the U.S. Environmental Protection Agency (EPA) is collecting ambient aqueous samples in the vicinity of abandoned, uncontrolled sites. These samples are analyzed by several laboratories throughout the United States. The results of these analyses could then be used by EPA to initiate the required litigation, clean-up activities, or other appropriate remedial measures.

As part of this larger program, Versar, a private laboratory, measures the concentration levels of specific metals and inorganic compounds for samples provided by EPA. The entire analysis which Versar performs includes the following: sample preparation, data reduction, designated quality assurance and quality control. As to specific chemicals, Versar measures the levels of twenty-four metals and six compounds (ammonia, cyanide, fluoride, hydrogen ion (pH), total organic carbon, and sulfides). By October 1981, Versar had analyzed about 1,000 different samples. Of these, about one hundred were split (analyzed in duplicate) so that an estimate of the total laboratory error could be made. The results of these analyses may be used by EPA in its enforcement and/or standard-setting functions.

Often, standards are set by comparing mean concentration levels, $\overline{C}$, of a hazardous compound or metal with some specified unsafe (or otherwise undesirable) level. Multimedia Environmental Goals (MEGs) for concentrations of compounds were established by the Environmental Protection Agency to aid in environmental assessment. They represent estimated levels of contaminants that (1)

0065-8812-82-6205-0220-$2.00

will not produce negative effects in thesurrounding populations or ecosystems or (2) represent control limits achievable through technology. Discharge MEGs (DMEGs) represent pollutant levels that appear not to adversely affect persons or ecosystems for short periods of time (less than 8 hours). The acceptability of each discharge may be determined by comparing the concentration for each constituent with the appropriate DMEG.

Another criterion often used is to compare the highest or the second highest concentration of the pollutant measured during a specified period at a monitoring site with a standard. Yet a third is to specify the percentage of data points or the expected proportion of time the specified unsafe level may be exceeded by the concentration, C', at the monitoring site. More specifically, the criterion could state that 95% of the hourly-averaged data have to be below a certain level (letting the firm exceed the standard approximately one hour per day).

The implementation of such criteria typically involves measurements which are subject to a number of sources of variability. One obvious source is the actual time and/or space variation in C. For example, the amount of silver in the waste water of a smelter clearly changes over time; the concentration of mercury in a lake clearly has natural spacial variation. In addition, the concentration, C, also depends on random fluctuations: the weather, the tides, the mineral content of the ore, etc. These kinds of variation cannot be eliminated, but they can be accounted for in a statistical model. Further, the very act of measuring C introduces additional variability: measurement errors may be introduced by the equipment or by the technicians performing the analysis; sampling errors may be introduced by the representation of a large body of water by only a small sample from it; selection error may arise by using inappropriate techniques for collecting or subsampling the primary sample. These errors can also be quantified and should be "eliminated" from the data by sound statistical techniques. Below, we will describe one stochastic model describing the actual concentration C and its noisy, measured version C', together with procedures for making statistical inferences regarding C. In particular, decision bases are needed for classifying sources as "clean," i.e., meeting the standards, or "dirty," i.e., not meeting the standards.

If we assumed that the measured concentration C' and the actual concentrations were normal, then standard statistical procedures would suffice. However, actual data tend to be lognormally distributed with multiplicative errors. Thus, for example, the average of the noisy process C' and the true process C are different! Since the standards stipulate the actual C and not C', models must be developed to estimate C from C'.

We shall limit our discussion here to a model appropriate to some of the Versar data described above. We consider the situation where the concentration levels of the material are always large enough to be above detection limits and systematic natural fluctuations in the levels (say, due to purely geographic considerations) can be neglected as a separate source of variation. Particular examples are the levels of aluminum, iron, or zinc.

For the cases mentioned above a one-way analysis of variance model would be appropriate. To compare two population means based on independent normal statistics, we may use the familiar t-test. Very often we have more than two populations of interest. A statistical procedure generalizing the t-test to this case is called ANalysis Of VAriance (ANOVA). It provides estimates for the variance between the populations (called the "between-variance") and for the common variance of the populations (called the "within-variance"). A comparison of these variances may be used to test the equality of the population means. In our example, EPA provides my firm with many samples from each abandoned, uncontrolled site. Each of these samples may then be split. The between-variance from ANOVA would then estimate the natural variation within the site while the within-variance would estimate the laboratory, measurement, etc. variation. The process C could then be estimated by "eliminating" this latter variation from the noisy data, C'.

In the section below we present the model in greater detail. We point out that our new procedure might show a firm to be "clean" whereas the usual models for the actual data would show it to be "dirty." Following this we present an example. Unfortunately EPA holds the proprietory rights on the data indicated above. Consequently, we discuss data from five textile mills given in Dunn, et al.[1]

THE MODEL

We assume that our measured process C' is a noisy version of the actual concentration C. In many important cases the models have errors proportional to the concentrations (or approximately so). Engineers tend to discuss errors in per cent and not in absolute terms. They naturally describe the error process E as multiplying the true process C to form the noisy process C', i.e. we assume C' = CE and C and E are independent. Since our data are well represented by the family of lognormal distributions, we assume X' = log C', X = log C, and e = log E are all normal random variables, that X' = X + e and that X and e are independent. We now have the necessary set-up for our ANOVA model. The process e will be assumed to have mean zero and variance equal to the "within variance" of ANOVA. The process X will be estimated by assuming it has mean equal to the average of X' and variance equal to the "between variance" of ANOVA. Once X and e are known, then C can be estimated.

More precisely now we suppose that C'_{jk}, j=1, .., J:k = 1, ..., K are the data for the jth sample and the kth split. The one-way ANOVA model is specified by

$$X'_{jk} = \mu + A_j + e_{jk} \qquad \begin{array}{l} j = 1, \ldots, J \\ k = 1, \ldots, K \end{array}$$

where:

μ is a constant

A_j is $N(0,\sigma^2)$ is the natural variation

e_{jk} is $N(0,\sigma_e^2)$ is the measurement variation

and the true transformed process is

$$X_{jk} = \mu + A_j.$$

The usual statistics from ANOVA are:

$\overline{X}'$ = average of the data

PSS = process sums-of-squares

$$= K\sum_j (A_j - \overline{X}')^2$$

WSS = within sums-of-squares

$$= \sum_j \sum_k (X'_{jk} - A_j)^2$$

where

A_j = average of the jth sample

$$= \sum_k X'_{jk}/K.$$

The standard estimators are:

$$\hat{\mu} = \overline{X}'$$

$$\hat{\sigma}^2 = \frac{PSS}{K(J-1)} - \frac{WSS}{JK(K-1)} = \text{between variance}$$

$$\hat{\sigma}_e^2 = WSS/J(K-1) = \text{within-variance.}$$

So now we "know" the processes X and e. We now need to estimate the process C = exp(X). This, unfortunately, is not as easy as it appears. As an example, consider the mean level m of C. It is

$$m = \exp\left(\mu + \frac{1}{2}\sigma^2\right).$$

However,

$$\text{Expected value } \left(\exp\left(\hat{\mu} + \frac{1}{2}\hat{\sigma}^2\right)\right) >$$

$$> \exp\left(\text{Expected value } \left(\hat{\mu} + \frac{1}{2}\hat{\sigma}^2\right)\right)$$

$$= \exp\left(\mu + \frac{1}{2}\sigma^2\right) = m,$$

by Jensen's Inequality. So the problem of estimating m is non-trivial! We shall use a function given by Finney (2)

$$g(x,\nu) = \sum_{n=0}^{\infty} \left(\frac{x}{4}\right)^n \frac{1}{n!} \frac{\Gamma\left(\frac{\nu}{2}\right)}{\Gamma\left(\frac{\nu}{2} + n\right)}.$$

Using this function, we can estimate m by $\hat{m}$ (see Dunn, et al.(1) or Gleit(3)):

$$\hat{m} = e^{\hat{\mu}}\, g\left(\frac{1}{K}\left(1 - \frac{1}{J}\right)PSS,\ J-1\right) g\left(-\frac{1}{K}WSS,\ J(K-1)\right).$$

$\hat{m}$ has the property that

$$\text{Expected value } (\hat{m}) = m.$$

It can be shown that

$$\hat{m} > \exp(\hat{\mu}).$$

Thus our estimator $\hat{m}$ is always larger than the exponential of the average value of the log of the data. This latter quantity is called the geometric mean of the data. So the reported concentration levels obtained by the geometric mean are biased downwards. Thus, it is possible that the true mean level m of C is above the standards set by the regulatory agencies whereas the geometric mean of the data might be below them. In other words, the procedure outlined above may show a firm to be "dirty" while the geometric mean of the data would show it to be "clean."

AN EXAMPLE (from Dunn et al.(1))

Data were collected on the discharge concentration of total suspended solids in the effluent of five textile mills manufacturing similar products. These data showed no temporal correlation and closely followed a lognormal frequency distribution. A model for the concentration levels was written as described above:

$$X'_{jk} = \mu + A_j + e_{jk}$$

where the subscript j refers to the process time at which the concentration level was measured and k refers to the sample split (or subsample number). The applicable standard at the time the data were collected involved the 95th percentile, $C_{.95}$, of the distribution of the actual concentration, C, as it compared to the standard of 300 $\mu g/m^3$.

There are (at least) three ways to proceed with the analysis:

1. assume the data are normal and use the usual theory,
2. assume the data are lognormal and use the usual lognormal theory (provided, say, in Johnson and Kotz(4)),
3. assume the data are lognormal and use the procedures outlined above for removal of measurement errors.

For the standard normal distribution we know that the 95th percentile is 1.645. Hence in (1) we would estimate $C_{.95}$ by

mean of C' + 1.645 std. deviation of C'.

For (2) we would estimate $C_{.95}$ by

exp(mean of X'+1.645 std. deviation of X').

The appropriate formula for (3) can be found in Dunn et al.(1) As the standard lognormal theory does not separate the variance components into true variance and measurement error, it must give a larger estimate of $C_{.95}$ than does our ANOVA method.

Estimates of $C_{.95}$, the 95th percentile of C

Textile Plant	Above ANOVA Model	Lognormal Theory	Normal Theory	# Data Points
Y	105.3	297.6	185.7	18
T	28.8	39.0	35.1	20
AA	133.2	154.5	143.1	30
Z	115.8	162.3	145.8	44
D	62.4	285.0	414.3	22

It can be seen that the estimates of $C_{.95}$ are all less than 300 $\mu g/m^3$ using our ANOVA model indicating that (based on our model) each of the five textile mills is "clean", i.e. not in violation of the law. The estimates using lognormal theory are very close to 300 for plants Y and D. In fact, a test of the hypothesis "$C_{.95} > 300$" would not be rejected for these two mills. Consequently they might be classified "dirty" if we used the standard lognormal model. Finally, using the standard normal model, $C_{.95}$ clearly exceeds 300 for plan D. One would surely classify it as "dirty"; in fact, the hypotheses "$C_{.95} < 300$" would be rejected with 99% confidence. Thus, mill D violates the law using standard normal statistics but is in compliance if we use the statistical procedures outlined above.

LITERATURE CITED

1. Dunn, J.E., Gleit, A., Leadbetter, M.R., and Manson, A.R., Estimation for log-normal data with multiplicative errors, 1981, working paper.

2. Finney, D.J., On the distribution of a variate whose logarithm is normally distributed, J. Royal Stat. Soc. (Series B) 7 (1941), 155-161.

3. Gleit, A., Estimation of functions of the parameters of a normal distribution, Comm. in Statistics (to appear).

4. Johnson, N. and Kotz, S., Continuous Univariate Distributions -1, Wiley, New York, 1970.

HISTORY OF CHEMICAL ENGINEERING

100 The History of Penicillin Production

ION EXCHANGE

14 Ion Exchange
24 Adsorption, Dialysis, and Ion Exchange
152 Adsorption and Ion Exchange
179 Adsorption and Ion Exchange Separations
219 Recent Advances in Adsorption and Ion Exchange

KINETICS

4 Reaction Kinetics and Transfer Process
25 Reaction Kinetics and Unit Operations
72 Recent Advances in Kinetics
73 Kinetics and Catalysis
83 Recent Advances in Kinetics and Catalysis

MATHEMATICS

31 Advances in Computational and Mathematical Techniques in Chemical Engineering
37 Applied Mathematics in Chemical Engineering
42 Statistical and Numerical Methods in Chemical Engineering
50 Optimization Techniques

MINERALS

15 Mineral Engineering Techniques
20 Liquid Metals Technology—Part I
43 Recent Advances in Ferrous Metallurgy
85 Fossil Hydrocarbon and Mineral Processing
173 Fundamental Aspects of Hydrometallurgical Processes
180 Spinning Wire from Molten Metals
216 Processing of Energy and Metallic Minerals

PETROCHEMICALS

34 Petrochemicals and Petroleum Refining
49 Polymer Processing
127 Declining Domestic Reserves—Effect on Petroleum and Petrochemical Industry
135 The Petroleum/Petrochemical Industry and the Ecological Challenge
142 Optimum Use of World Petroleum
212 Interfacial Phenomena in Enhanced Oil Recovery

PETROLEUM PROCESSING

34 Petrochemicals and Petroleum Refining
54 Hydrocarbons from Oil Shale, Oil Sands, and Coal
98 Methanol Technology and Economics
103 C_4 Hydrocarbon Production and Distribution
127 Declining Domestic Reserves—Effect on Petroleum and Petrochemical Industry
135 The Petroleum/Petrochemical Industry and the Ecological Challenge
142 Optimum Use of World Petroleum
155 Oil Shale and Tar Sands

PHASE EQUILIBRIA

2 Pittsburgh and Houston
3 Minneapolis and Columbus
6 Collected Research Papers
81 Phase Equilibria and Related Properties
88 Phase Equilibria and Gas Mixtures Properties

PROCESS DYNAMICS

36 Process Dynamics and Control
46 Process Systems Engineering
55 Process Control and Applied Mathematics
159 Chemical Process Control
214 Selected Topics on Computer-Aided Process Design and Analysis

SEPARATION

91 Unusual Methods of Separation
120 Recent Advances in Separation Techniques
192 Recent Advances in Separation Techniques—II

SONICS

1 Ultrasonics—Two Symposia
109 Sonochemical Engineering

THERMODYNAMICS

7 Applied Thermodynamics
44 Thermodynamics
140 Thermodynamics-Data and Correlations

TRANSPORT PROPERTIES

16 Mass Transfer
56 Selected Topics in Transport Phenomena
77 Fundamental Research on Heat and Mass Transfer

MISCELLANEOUS

26 Chemical Engineering Education—Academic and Industrial
48 Chemical Engineering Reviews
70 Small-Scale Equipment for Chemical Engineering Laboratories
76 High Pressure Technology
112 Engineering, Chemistry, and Use of Plasma Reactors
125 Vacuum Technology at Low Temperatures
143 Standardization of Catalyst Test Methods
160 Continuous Polymerization Reactors
182 Biorheology
183 The Modern Undergradate Laboratory Innovative Techniques
185 Electro organic Synthesis Technology
186 Plasma Chemical Processing
187 Chronic Replacement of Kidney Function
194 Hazardous Chemical—Spills and Waterborne Transportation
203 A Review of AIChE's Design Institute for Physical Property Data (DIPPR) and Worldwide Affiliated Activities
204 Tutorial Lectures in Electrochemical Engineering and Technology
206 Controlled Release Systems
217 New Composite Materials and Technology
220 Uncertainty Analysis for Engineers

MONOGRAPH SERIES

1 Reaction Kinetics by Olaf Hougen
2 Atomization and Spray Drying by W.R. Marshall, Jr.
3 The Manufacture of Nitric Acid by the Oxidation of Ammonia—The DuPont Pressure Process by Thomas H. Chilton
4 Experiences and Experiments with Process Dynamics by Joel O. Hougen
5 Present Past, and Future Property Estimation Techniques by Robert C. Reid
6 Catalysts and Reactors by James Wei
7 The 'Calculated' Loss-of-Coolant Accident by L.J. Ybarrondo, C.W. Solbrig, H.S. Isbin
8 Understanding and Conceiving Chemical Process by C. Judson King
9 Ecosystem Technology: Theory and Practice by Aaron J. Teller
10 Fundamentals of Fire and Explosion by Daniel R. Stull
11 Lumps, Models and Kinetics in Practice by Vern W. Weekman, Jr.
12 Lectures in Atmospheric Chemistry by John H. Seinfeld
13 Advanced Process Engineering by James R. Fair
14 Synfuels from Coal by Bernard S. Lee